LOGARITHMS

LOGARITHMS

ITT FEDERAL ELECTRIC CORPORATION
A Subsidiary of International Telephone & Telegraph Corp.

John Wiley & Sons, Inc.
New York • London • Sydney • Toronto

Library of Congress Catalog Card Number: 77-38626

ISBN 0-471-25683-8

Printed in the United States of America.

10 9 8 7 6 5 4 3 2 1

Preface

Logarithms are useful to anyone who consistently works with very large or very small numbers and involved computations, whether in mathematics or science, industry or research. Logarithms were developed during the seventeenth century when people needed a faster and less tedious method of making complicated calculations involving multiplications, divisions, powers, and roots. Today, of course, we have calculating machines and computers to perform many complicated operations; however, they are not available to everyone. Hence logarithms are used.

Many students find logarithms necessary for their science and mathematics courses, but often they either never learned or have forgotten how to use them. With this program they can teach themselves how to use logarithms, or review their knowledge. Of course teachers who use this program with their classes will have additional time for teaching more difficult material. Logarithms also have applications in such diverse fields as banking, chemistry, civil service, electronics, insurance, physics, statistics, and engineering. Anyone who finds slide-rule accuracy inadequate, but who is not provided with a desk computer, will find logarithms the most convenient way to deal with involved computations.

Logarithms is presented so that you can learn logically rather than mechanically. It stresses ability to actually use logarithms; with this common-sense approach you should always be able to reconstruct rather than merely recall how to use logarithms, even if you have not used them for some time. And it is self-contained; all tables are included in the book. *Logarithms* will give you a working knowledge of an important mathematical tool.

This project was under the direction of Oswald Wolf, M. S., Columbia University, Manager of Special Services Department of the Federal Electric Company. Henry Shleuder, Training Manager, is the author of the text. The author's thanks go to other members of the Training Branch for their guidance and suggestions. He is especially grateful to Mr. Joe Di Mauro for his invaluable editorial assistance and to Mrs. Carole Wells for her patience and understanding in typing the difficult original manuscript.

FEDERAL ELECTRIC CORPORATION

Paramus, New Jersey
November 1971

Prerequisites

If you have completed a course in elementary algebra, or the equivalent, you should have all the background you need for *Logarithms*. Specifically, you should be able to

(1) perform the simple arithmetic operations of addition, subtraction, multiplication, and division;

(2) solve proportionality problems such as $\frac{a}{b} = \frac{c}{d}$;

(3) visualize quantities represented by simple literal expressions such as

$$\begin{aligned} A &= \text{area}, \\ t &= \text{radius of a circle}, \\ x &= \text{unknown}, \\ \log x &= \text{power of the number 10}; \end{aligned}$$

(4) solve simple linear equations in one unknown by performing the same operation to each side of the equation:

EXAMPLE: $3x = 120$
dividing each member by 3,
$x = 40$;

(5) consistently read a numerical table correctly after several practice readings;

(6) define the words "root" and "power" and give examples of each.

Objectives for Logarithms

Main Objective

After completing this book you will be able to perform the following mathematical operations with the use of logarithms:

(a) multiplication,
(b) division,
(c) raising a number to a power,
(d) extracting roots.

UNIT 1

You will be able to perform multiplications and divisions of exponential numbers with the base of 10.

UNIT 2

You will be able to operate with numbers raised to zero power.

UNIT 3

You will be able

(a) to convert an exponential number into whole logs;
(b) to recognize numbers that, if converted, would become fractional logs.

UNIT 4

You will be able

(a) to use a simplified log table;
(b) to determine the characteristic of a number;
(c) to find the mantissa in a simplified log table.

UNIT 5

You will be able

(a) to distinguish between characteristics and mantissas;
(b) to relate exponential numbers to logs.

UNIT 6

You will be able to use a representative log table

(a) to find the log for numbers with three significant digits;
(b) to interpolate the fourth significant digit;
(c) to use proportional parts in the log table.

UNIT 7

You will be able to find negative logs by a logical method

(a) to convert a fractional number into a log;
(b) to use three versions of negative logs;
(c) to change from one version to another by a logical process.

UNIT 8

You will be able to find the antilog of mantissas

(a) with four places to the nearest figure in the log table;
(b) to interpolate the remainder;
(c) to use proportional parts from the log table.

UNIT 9

You will be able to find antilogs consisting of characteristics and mantissas

(a) to relate characteristics to exponential numbers with the base of 10;
(b) to reconvert a negative log to a fractional number and then to a decimal fraction.

UNIT 10

You will be able to perform multiplications of fractional numbers by the use of logs

(a) to relate logs to exponential numbers (this method will ensure greater retention and easier recall after an extended period of disuse);
(b) to perform multiplications of numbers which result in positive and negative characteristics.

UNIT 11

You will be able to perform divisions by the use of logs

(a) to relate logs to exponential numbers;
(b) to perform divisions of numbers which result in positive and negative characteristics.

UNIT 12

You will be able to raise numbers to a power by the use of logs

(a) to advance your skill to the application of raising a number to a power, building on the knowledge you acquired in preceding units;
(b) to perform power operations in a simple manner instead of an involved and tedious procedure of the longhand method.

UNIT 13

You will be able to perform extraction of roots

(a) as an extension of the logical process learned in preceding sets;
(b) by the simple method of using logs instead of the tedious method of longhand calculations.

UNIT 14

You will be able to change negative logs to any convenient equivalent form without changing the value.

UNIT 15

You will be able to use cologs when working with involved calculations.

UNIT 16

You will be able to use the natural logarithm system.

How to Use This Book

With this self-teaching book you can learn logarithms at whatever pace is best for you. Throughout the program you will be actively participating—answering questions, completing calculations, and using logs—and you will find out immediately whether your answers were right or wrong. This programmed instruction book, which consists of 16 units, is divided into numbered frames; each frame presents some new information and asks you a question. After you have answered, check the answer given below the dotted line. If your answer does not agree with the one given, check your arithmetic. If the answer still does not agree, or if you are unsure about the method used, review the preceding frames before going on. Thus you always move at your own speed, going on only after you are sure you understand the material. A card on which the tables of common and natural logarithms are printed is inserted inside the front cover. You may wish to use this card to mask the answers as you work through the program.

The objectives on the preceding pages specify what you should be able to do at each point in the program. The self-test at the end of each unit will show whether you have achieved those objectives. After you complete each test, check you answers; as always, if any are incorrect, be sure you understand why before going on. A final self-test is provided at the end of the book.

Contents

UNITS

1 Exponential Numbers with Base 10 1

2 Numbers Raised to Zero Power 10

3 Logarithms of Multiples of 10 and Numbers 1 to 10 13

4 How to Use a Simplified Logarithm Table 21

5 More About Characteristics and Mantissas 32

6 How to Use an Expanded Logarithm Table 41

7 Negative Logarithms 59

8 Finding the Antilog of Mantissas 71

9 Finding the Antilog of Characteristics and Mantissas 87

10 Multiplication by the Use of Logarithms 98

11 Division by the Use of Logarithms 112

12 Raising to a Power by the Use of Logarithms 123

13 Extracting Roots by the Use of Logarithms 131

14 One More Way to Write Negative Logarithms 157

15 Cologs and Involved Problems 163

16 Natural Logarithms 180

Final Self-Test 197

Common Logarithms (including proportional parts) 213

Natural Logarithms (including mean differences and natural logarithms of 10^{+n} and 10^{-n}) 214

LOGARITHMS

UNIT ONE

Exponential Numbers with Base 10

1.

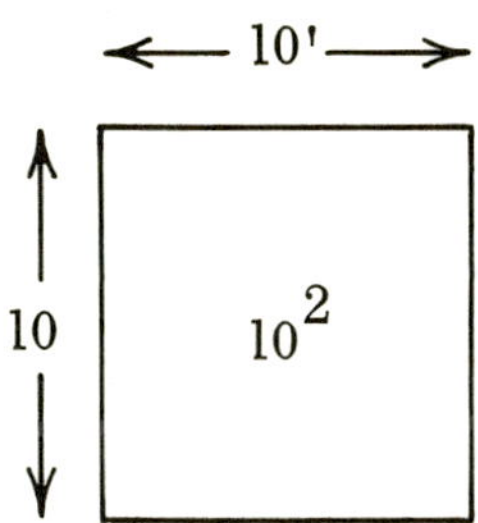

A square with sides of 10 ft can be described as __10__ x __10__ or __10^2__.

- -

10

10

10^2

2. A cube with side dimensions of 10 can be described as 10 x 10 x 10 or __10^3__.

- -

10^3

3. If 10^3 can be visualized as a cube with side dimensions of 10 (feet, inches, etc.), would you have difficulty in visualizing 10^4 or, for that matter, 10^5 or 10^6?

- - - - - - - yes - - - - - - - - - - - -

You probably would. Whereas 10^2 and 10^3 can be shown geometrically, 10^4, 10^5, 10^6, etc., are strictly mathematical terms.

4. The symbol 10^2 means multiply 10 x 10. The symbol for the multiplication 10 x 10 x 10 would be 10^3.

- - - - - - - - - - - - - - - - - -

10^3

5. One thousand can be shown as 1000, 10 x 10 x 10 or 10^3.

- - - - - - - - - - - - - - - - - -

10^3

6. 10 x 10 x 10 x 10 = 10^4

- - - - - - - - - - - - - - - - - -

10^4

7. 10^2 = 100
10^3 = 1000
10^4 = 10,000
10^5 = 100000
10^6 = 1,000,000
10^7 = 10, 000, 000

- - - - - - - - - - - - - - - - - -

10, 000

100, 000

1, 000, 000

8. Instead of using the long number 100,000,000, we may write it in mathematical shorthand as ____.

- - - - - - - - - - - - - - - - - - -

10^8

9. A number such as 10^2 is called an exponential number and consists of two parts: 10 is the base and 2 is the exponent. A number with a base of 10 and exponent of 10 would be written as ____.

- - - - - - - - - - - - - - - - - - -

10^{10}

10. In 10^5 the exponent of the base 10 is ____.

- - - - - - - - - - - - - - - - - - -

5

11. The exponent in 10^6 is 6. What is the base? ____

- - - - - - - - - - - - - - - - - - -

10

12. The exponents in 10^2, 10^3, 10^4, 10^5, 10^6, and 10^7 are 2, 3, ____, ____, ____, and ____, respectively.

- - - - - - - - - - - - - - - - - - -

4

5

6

7

13. The base in 10^2, 10^3, 10^4, 10^5, 10^6, and 10^7 is ______.

10

14. 10^3 means the same as 10 x 10 x 10. 10 x 10 x 10 = 1000. Therefore 10^3 means the same as 1000. Expressed in commom mathematical form:

$10^3 = 10 \times 10 \times 10 = 1000$

The base 10 appears ______ times in the multiplication 10 x 10 x 10 .

3

15. If in the expression 10^3 (visualized as a multiplication) the base 10 appears 3 times, and in the expression 10^2 the base 10 appears 2 times, then in the expression 10^1 the base 10 will appear ______ times(s).

1

16. In the mathematical expression 10^1, 10 is the ______ and 1 is the ______.

---

base

exponent

17. In multiplying exponential numbers having a common base, the base remains the same and the exponents are added.

$10^3 \times 10^2 = 10^{3+2} = 10^5$

$10^7 \times 10^6 = 10^{7+6} = 10^{13}$

$10^1 \times 10^4 = \underline{10^{1+4}} = \underline{10^5}$

- - - - - - - - - - - - - - - - - - -

$10^{1+4} = 10^5$

18. If there is any doubt in your mind, the following examples will serve to dissipate it:

$3^2 \times 4^2$ does not equal ($3^4 = \underline{81}$) or ($4^4 = \underline{256}$).

Converting the two terms into numbers and then multiplying them is as follows:

$3^2 \times 4^2 = 9 \times 16 = \underline{144}$.

Compare the underlined figures and draw your own conclusions.

19. $10^2 \times 10^2 = 10^{2+2} = \underline{10^4}$

- - - - - - - - - - - - - - - - - - -

10^4

20. $10^1 \times 10^1 = \underline{10^2}$

$10^3 \times 10^3 = \underline{10^6}$

$10^6 \times 10^1 = \underline{10^7}$

- - - - - - - - - - - - - - - - - - -

10^2

10^6

10^7

21. Not only do we multiply exponential numbers by adding the exponents but we divide exponential numbers by subtracting exponents, provided they have the same base; for instance,

$$\frac{10^6}{10^3} = 10^{6-3} = 10^3$$

$$\frac{10^7}{10^2} = 10^{7-2} = 10^5$$

$$\frac{10^8}{10^4} = 10^{8-4} = \underline{10^4}$$

- - - - - - - - - - - - - - - - - - - -

10^4

22. $\frac{10^7}{10^3} = \underline{10^4}$ $\frac{10^{18}}{10^2} = \underline{10^{16}}$ $\frac{10^6}{10^3} = \underline{10^3}$ $\frac{10^4}{10^3} = \underline{10^1}$

- - - - - - - - - - - - - - - - - - - -

10^4

10^{16}

10^3

10^1

23. To divide exponential numbers having the same base, we <u>subtract</u> (add, subtract) exponents.

- - - - - - - - - - - - - - - - - - - -

subtract

24. To multiply exponential numbers having the same base, we ________ (add, subtract) exponents.

- - - - - - - - - - - - - - - - - -

add

25. The exponent denotes how often the base would appear if the number were written as a common mathematical operation: 10^2 written as a common mathematical operation is 10 x 10. It appears twice. Similarly, 10^3 is written as a common mathematical operation in the form ________________.

- - - - - - - - - - - - - - - - - -

10 x 10 x 10

26. The exponetial form for 10 is _____.

- - - - - - - - - - - - - - - - - -

10^1

27. Simply expressed, you count how often the number appears in a common mathematical operation. The count becomes the exponent. In 10^4 the base is _____ and the exponent is _____.

- - - - - - - - - - - - - - - - - -

10

4

28. In common usage the exponent is placed a little higher and to the right of the __base__ and denotes how often the __base__ (base, exponent) appears in a common mathematical operation or number.

- - - - - - - - - - - - - - - - - -

base

base

29. To multiply or divide exponential numbers, their __base__ (base, exponent) must be common.

- - - - - - - - - - - - - - - - - -

base

SELF-TEST

1. Identify the base and the exponent in the expression 10^6.

base – 10 exponent 6

2. Perform the following operations:

(a) $10^5 \times 10^3 \times 10^6 \times 10^2$ = 10^{16}

(b) $10 \times 10 \times 10^3 \times 10^1$ = 10^6

(c) $10^6 \div 10^3 = \frac{10^6}{10^3}$ = 10^3

(d) $10^{15} \div 10^5$ = 10^{10}

(e) $\frac{10^{17}}{10^7}$ = 10^{10}

(f) $\frac{10^8}{10^3}$ = 10^5

(g) $\dfrac{10^5}{10^2}$ = 10^3

Answers

1. 10 is the base and 6 is the exponent.

2. (a) 10^{16}

   (b) 10^6

   (c) 10^3

   (d) 10^{10}

   (e) 10^{10}

   (f) 10^5

   (g) 10^3

UNIT TWO

Numbers Raised to Zero Power

The statement $10^0 = 1$ is correct, but how can we be sure? The rules that we have learned so far do not apply to this case. There must be a different approach, and there is.

1. $\frac{10,000}{10,000}$ = ______ $\frac{1000}{1000}$ = ______ $\frac{100}{100}$ = ______ $\frac{10}{10}$ = ______

- - - - - - - - - - - - - - - - - - - -

1

1

1

1

2. $\frac{x}{x}$ = ______

- - - - - - - - - - - - - - - - - - - -

1

3. Any number (except zero) divided by itself equals ______.

- - - - - - - - - - - - - - - - - - - -

1

4. We have learned that to divide exponential numbers we have only to subtract one exponent from the other; for instance;

$$\frac{10^4}{10^4} = 10^{4-4} = 10^0$$

- -

$\frac{10^4}{10^4}$ is the same as $10^{4-4} = 10^0$: $\frac{10^4}{10^4}$, being a number divided by itself, is equal to __1__; therefore 10^0 is also equal to __1__.

- -

1

1

5. The following example will further prove and illustrate that 10^0 is equal to __1__.

$$\frac{10^3}{10^3} = \underline{\quad 1 \quad}$$

$$\frac{10^3}{10^3} = 10^{3-3} = 10^0 = \underline{\quad 1 \quad}$$

6. In our further work we need not concern ourselves with how these calculations were obtained. We should just remember that 10^0 = __1__.

- -

1

SELF-TEST

Show that $10^0 = 1$

1. $\frac{10^5}{10^5} = 1$; $\frac{10^5}{10^5} = 10^{5-5} = 10^{0} = 1$

2. $\frac{10^7}{10^7} = 1$; $\frac{10^7}{10^7} = 10^{7-7} = 10^0 = 1$

3. $\frac{10^{17}}{10^{17}} = 1$; $\frac{10^{17}}{10^{17}} = 10^{17-17} = 10^0 = 1$

4. $\frac{x}{x} = 1$

5. Any number, except 0, divided by itself equals 1.

Answers

1. $\frac{10^5}{10^5} = 1$; $\frac{10^5}{10^5} = 10^{5-5} = 10^0 = 1$

2. $\frac{10^7}{10^7} = 1$; $\frac{10^7}{10^7} = 10^{7-7} = 10^0 = 1$

3. $\frac{10^{17}}{10^{17}} = 1$; $\frac{10^{17}}{10^{17}} = 10^{17-17} = 10^0 = 1$

4. $\frac{x}{x} = 1$

5. 1

UNIT THREE

Logarithms of Multiples of 10 and Numbers of 1 to 10

1. $10^4 = 10,000$

$10^3 = 1000$

$10^2 =$ _____

$10^1 =$ _____

$10^0 =$ _____

- - - - - - - - - - - - - - - - - -

100

10

1

2. In all our examples we used a common base which was _____.

- - - - - - - - - - - - - - - - - -

10

3. A different form of exponential numbers goes under the name "logarithm." If you see the term, "log 4," read it "log of four." Similarly, "log 6" is read "log of 6," etc.

For most of this book we are concerned with logarithms to the base 10, called common logarithms. (In Unit 16 we use natural logarithms.) For better identification you will see logs with the subscript "10" in the following form:

$\log_{10} 10,000$ or $\log_{10} 10^4 = 4$

$\log_{10} 1000$ or $\log_{10} 10^{3} =$ 3

$\log_{10} 100$ or $\log_{10} 10^{2} =$ 2

$\log_{10} 10$ or $\log_{10} 10^{1} =$ 1

$\log_{10} 1$ or $\log_{10} 10^0 = 0$

- - - - - - - - - - - - - - - - - -

$\log_{10} 10^3 = 3$

$\log_{10} 10^2 = 2$

$\log_{10} 10^1 = 1$

4. In our examples the base is always 10. Since it is repetitive, we may take the base 10 for granted and drop it. Below are a number of comparisons that speak for themselves:

$\log 10,000$ or $\log 10^4 = 4$

$\log 1000$ or $\log 10^3 = 3$

$\log 100$ or $\log 10^2 =$ 2

$\log 10$ or $\log 10^1 =$ 1

- - - - - - - - - - - - - - - - - -

2

1

5. $\log 1$ or $\log 10^0 =$ 0

- - - - - - - - - - - - - - - - - -

0

6. For numbers 1, 10, 100, 1000, etc., we can find the log easily. We write or visualize it as a power to the base 10, such as 10^5, and just drop the __________ (base 10, exponent 5).

- - - - - - - - - - - - - - - - - -

base 10

7. What about the numbers 1 to 10?

log 1 = ______

log 2 = (we don't know yet)

log 10 = ______

- - - - - - - - - - - - - - - - - -

0

1

8. The number 2 is larger than 1. Therefore, if the log 1 = 0, the log 2 must be ________ (greater, smaller) than 0.

- - - - - - - - - - - - - - - - - -

greater

9. The number 2 is smaller than 10. Therefore, if the log 10 = 1, the log 2 must be ________ (greater, smaller) than 1.

- - - - - - - - - - - - - - - - - -

smaller

10. The log 2 lies between 0 and 1. Therefore the log 2 must be a ______________ (whole, fractional) number.

\- - - - - - - - - - - - - - - - - - -

fractional

11. If log 2 is a fractional number, what would log 3 be? ______________ (whole number, fraction).

\- - - - - - - - - - - - - - - - - - -

fraction

12. The log 7 is a __________.

\- - - - - - - - - - - - - - - - - - -

fraction

13. The log 10 = 1 and is a ______________.

\- - - - - - - - - - - - - - - - - - -

whole number

Note: Any log that is a fraction is always written as a decimal fraction. The log 2 = 0.3. The log 3 = 0.4771. (Taken from a log table.)

14. Reverting to the exponential form, log 2 = 0.3 = $\log 10^{0.3}$ and log 3 = 0.4771 = __________.

\- - - - - - - - - - - - - - - - - - -

$\log 10^{0.4771}$

15.

Exponential Form	Log Form
$200 = 100 \times 2$	$\log 100 = 2$
$= 10^{2} \times 2$	$\log 2 = 0.3$
$= 10^{2} \times 10^{0.3}$	$\log 200 = 2 + 0.3 =$ ______
$= 10^{2+0.3} =$ ______	

- - - - - - - - - - - - - - - - - - -

$10^{2.3}$

2.3

16. In exponential form a multiplication is reduced to an addition of exponents. Since logarithmic numbers are a form of exponential numbers, multiplications ________ (will, will not) become simple additions.

- - - - - - - - - - - - - - - - - - -

will

17. If, in logarithms, multiplications are reduced to additions, it follows with mathematical logic that divisions will be reduced to ____________.

- - - - - - - - - - - - - - - - - - -

subtractions

18. Logarithms are a simplifed form of exponetial numbers in which the ________ (base, exponent) is taken for granted and left out.

- - - - - - - - - - - - - - - - - - -

base

19. Multiplications can be performed by determining the logarithms of the common numbers and then ______________ (adding, subtracting) the logarithms.

- - - - - - - - - - - - - - - - - -

adding

20. Divisions can be performed by determining the logarithms of the common numbers and then ______________ (adding, subtracting) the logarithms.

- - - - - - - - - - - - - - - - - -

subtracting

21. 2000 = 1000 x 2

4000 = 1000 x ______

log 2000 = log 1000 + log 2

log 4000 = log 1000 + log ______

log 2000 = log 10^3 + $\log^2$

log 4000 = log 10^3 + log ______

The log 10^3 is simply ______.

log 10^4 = ______

- - - - - - - - - - - - - - - - - -

4

4

4

3

4

22. Log 10, log 100, log 1000, etc., can be found by simply writing the common numbers in their exponential form and just dropping the base of 10, thus:

10	= 10^1	log 10	= 1
100	= 10^2	log 100	= 2
10,000	= 10^4	log 10,000	= 4
100,000	= 10^5	log 100,000	= 5
1000	= 10^3	log 1000	= 3
1,000,000	= 10^6	log 1,000,000	= 6

- - - - - - - - - - - - - - - - - - -

10^4, 4

10^5, 5

10^3, 3

SELF-TEST

1. log 10^7 = 7
2. log 10^5 = 5
3. log 10^0 = 0
4. log 10^{11} = 11
5. log 10^1 = 1
6. log 10,000 = 4
7. log 1,000,000 = 6

8. Log 3 will be a whole or fractional number?

9. $\log 10^{0.03}$ = ______

10. $\log 10^{0.4771}$ = ______

Answers

1. 7
2. 5
3. 0
4. 11
5. 1
6. 4
7. 6
8. Fractional number
9. 0.03
10. 0.4771

UNIT FOUR

How To Use a Simplified Logarithm Table

1. Log 4000 can be broken down into log 1000 + log 4. Log 1000 is simply ______.

- -

3

Logarithm Table

N	log	N	log	N	log	N	log	N	log
1	0000	**21**	3222	**41**	6128	**61**	7853	**81**	9085
2	3010	22	3424	42	6232	62	7924	82	9138
3	4771	23	3617	43	6335	63	7993	83	9191
4	6021	24	3802	44	6435	64	8062	84	9243
5	6990	25	3979	45	6532	65	8129	85	9294
6	7782	**26**	4150	**46**	6628	**66**	8195	**86**	9345
7	8451	27	4314	47	6721	67	8261	87	9395
8	9031	28	4472	48	6812	68	8325	88	9445
9	9542	29	4624	49	6902	69	8388	89	9494
10	0000	30	4771	50	6990	70	8451	90	9542
11	0414	**31**	4914	**51**	7076	**71**	8513	**91**	9590
12	0792	32	5051	52	7160	72	8573	92	9638
13	1139	33	5185	53	7243	73	8633	93	9685
14	1461	34	5315	54	7324	74	8692	94	9731
15	1761	35	5441	55	7404	75	8751	95	9777
16	2041	**36**	5563	**56**	7482	**76**	8808	**96**	9823
17	2304	37	5682	57	7559	77	8865	97	9868
18	2553	38	5798	58	7634	78	8921	98	9912
19	2788	39	5911	59	7709	79	8976	99	9956
20	3010	40	6021	60	7782	80	9031		
N	**log**	**N**	**log**	**N**	**log**	**N**	**log**	**N**	**log**

Figure 1

2. For the time being we will not concern ourselves further with log 1000, but concentrate on logs of numbers 1 to but not including 10. For log 4 we consult a logarithm table. Figure 1 is a logarithm table. The first column contains the common numbers. Next to them are the logs of those numbers; for instance, the common number is 4. The log 4 is shown as 6021 and is written .6021. (Decimal points are not shown in the table and must be inserted.) If the common number is 3, the log 3 is .4771. If the common number is 6, the log 6 is .7782.

- - - - - - - - - - - - - - - - - -

.7782

3. If the common number is 8, the log 8 = .9031.

- - - - - - - - - - - - - - - - - -

.9031

4. The logarithm table (Figure 1) gives the logs of numbers 1 to 99 with decimal points omitted. Logs of 10, 100, 1000, etc., can be found by first determining their exponential form; we then simply drop the base of 10 and use the exponent only.

log 10 = 10^1 = 1

log 100 = 10^2 = 2

log 1000 = 10^3 = 3

log 10,000 = 10^4 = 4

- - - - - - - - - - - - - - - - - -

2

3

5. The logs of numbers 10, 100, 1000, etc., are called the characteristics of the numbers. Thus the characteristics of

log 10 = 1

log 100 = ____

log 1000 = ____

log 10,000 = ____

log 100,000 = ____

log 1,000,000 = ____

log 10,000,000 = 7

- - - - - - - - - - - - - - - - - - -

2

3

4

5

6

6. The characteristic of a number is always an integer (whole number). Therefore the characteristics of log 10, log 100, log 1000, etc., are ________ (integers, fractions).

- - - - - - - - - - - - - - - - - - -

integers

7. The characteristic of a number is always an integer. However, the logarithms of numbers greater than 1 but less than 10 are fractions. By consulting the log table we find the log of 5 is .6990. Knowing that all logs of numbers greater than 1 but less than 10 are fractions, we write

log 5 = .6990.

The log 9 = .9542. Therefore log 9 is a(n) fraction (integer, fraction).

- - - - - - - - - - - - - - - - - -

fraction

8. The log 2 appears in the log table as 3010 and we write it as .3010. By the same token

log 3 = .4771

log 4 = .6021

log 5 = .6990

log 6 = .7782

log 7 = .8451

log 8 = .9031

log 9 = .9542

- - - - - - - - - - - - - - - - - -

.4771

.6021

.6990

.7782

.8451

.9031

.9542

9. The fraction in a log has been given the name "mantissa." A mantissa is always a fraction. Log tables, however, do not show mantissas in the form of fractions. When using log tables the decimal point must be inserted.

The log table shows only mantissas but not characteristics. The characteristics of numbers ___will not___ (will, will not) appear in log tables.

- - - - - - - - - - - - - - - - - - - -

will not

10. $300 = 100 \times 3$

$300 = 10^2 \times$ ___3___

- - - - - - - - - - - - - - - - - - - -

3

11. $\log 10^2 = 2$

$\log 3 = .4771$

$\log 300 = \log 10^2 + \log 3 =$ ___2.4771___

- - - - - - - - - - - - - - - - - - - -

2.4771

12. In log 300 = 2.4771, 2 is the characteristic and .4771 is the ________________.

- - - - - - - - - - - - - - - - - -

mantissa

13. The characteristic ____________ (can, cannot) be found in log tables.

- - - - - - - - - - - - - - - - - -

cannot

14. The mantissa ____________ (can, cannot) be found in log tables.

- - - - - - - - - - - - - - - - - -

can

15. The characteristic must be found by writing the number (actually or mentally) in exponential form; 400 can be broken down into $10^2 \times 4$. Log 10^2 is simply ______.

- - - - - - - - - - - - - - - - - -

2

16. Log 4 is ____________.

- - - - - - - - - - - - - - - - - -

.6021

17. The characteristic of log 400 is ______ and the mantissa is ________. The log 400 = ___________.

2

.6021

2.6021

18. The common number 5000 can be broken down into 1000 x 5 = 10^3 x 5. Log 5000 = ____.6990.

- - - - - - - - - - - - - - - - - - - -

3

19. 60,000 can be broken down into 10,000 x 6 = 10^4 x 6; log 60,000 = ________.

- - - - - - - - - - - - - - - - - - - -

4.7782

20. 7,000,000 can be broken down into ________ x ____ = ____ x ____; log 7,000,000 = ________.

- - - - - - - - - - - - - - - - - - - -

1,000,000

7

10^6

7

6.8451

21. Inserting the decimal point in the mantissa for log 3, we would read ________.

- - - - - - - - - - - - - - - - - - - -

.4771

22. A decimal point must precede the mantissa figure shown in the log table. Therefore the actual mantissa for log 9 is ________.

.9542

23. To determine the log 500 we break the number down into $100 \times 5 = 10^2 \times 5$. The characteristic is the exponent and the mantissa is taken from a log table; log 500 = ________.

2.6990

24. The characteristic is always a(n) ________ (integer, fraction) and the mantissa is always a(n) ________ (integer, fraction).

---

integer

fraction

25. The characteristic represents the number 10 or multiples of 10. It is simply the power to which 10 is raised. The characteristic is always a(n) ________. The mantissa, which is always a(n) ________, represents numbers greater than 1 but less than ________.

---

integer

fraction

10

26. To find the log of a common number it is changed to multiplication consisting of an exponential number with the base of ____ and a number from 1 to, but not including, 10. The exponential number is used to determine the ____________ and the other number is used to find the ____________ in a log table.

- - - - - - - - - - - - - - - - - -

10

characteristic

mantissa

27. This is the log of a number: 13.9031. The mantissa, which is is a fraction, represents a number that is greater than 1 but less than 10; 13 is the ____________ and .9031 is the ____________.

- - - - - - - - - - - - - - - - - -

characteristic

mantissa

28. This is the log of a number: 4.6321. The characteristic is ______ and the mantissa is ________.

- - - - - - - - - - - - - - - - - -

4

.6321

29. The exponent of the exponential number with the base of 10 becomes the ____________ (characteristic, mantissa), and the number greater than 1 but less than 10 is used to find the ____________ (characteristic, mantissa) in a log table or by some other means.

- - - - - - - - - - - - - - - - - -

characteristic

mantissa

SELF-TEST

1. log 10,000 = 4
2. log 60,000 = $6 \times 10^4 = 4.7782$
3. log 5000 = $5 \times 10^3 = 3.6990$
4. log 60 = $6 \times 10^1 = 1.7782$
5. log 3 = .4771
6. log 900,000 = $9 \times 10^5 = 5.9542$
7. log 3,000 = $3 \times 10^3 = 3.4771$
8. log 400 = $4 \times 10^2 = 2.$
9. log 7 = .8451
10. log 1 = 0
11. log 10 = 1
12. log 20 = $2 \times 10^1 = 1.3010$
13. log 20,000 = $2 \times 10^4 = 4.3010$

Find the logs for the following expressions:

14. $10^5 \times 8$ 5.9031
15. $10^{11} \times 4$ 11.6021
16. $10^9 \times 4$ 9.6021

17. $10^7 \times 5$

18. $10^{15} \times 5$

19. $10^1 \times 3$

20. $10^2 \times 2$

Answers

1. 4
2. 4.7782
3. 3.6990
4. 1.7782
5. .4771
6. 5.9542
7. 3.4771
8. 2.6021
9. .8451
10. 0
11. 1
12. 1.3010
13. 4.3010
14. 5.9031
15. 11.6021
16. 9.6021
17. 7.6990
18. 15.6990
19. 1.4771
20. 2.3010

UNIT FIVE

More About Characteristics and Mantissas

1. 4000 can be broken down into $1000 \times 4 = 10^3 \times 4$. In log form

$$\log 4000 = \log 1000 + \log 4 = \log 10^3 + \log^4$$

$40 = 10 \times 4 = 10^1 \times 4$
In log form $\log 40 = \log 10 + \log 4 = \log 10^1 = \log 4$.

$400 = 100 \times 4 = 10^2 \times 4$
In log form $\log 400 = \log 100 + \log 4 =$ _____ $+ \log 4$.

- - - - - - - - - - - - - - - - -

10^2

2. The first parts of the mathematical expressions

$\log 10^1 + \log 4$

$\log 10^2 + \log 4$

$\log 10^3 + \log 4$

$\log 10^4 + \log 4$

are used to determine the characteristics of the logs. The characteristics 1, 2, 3, and 4 are simply the ____________ (bases, exponents).

- - - - - - - - - - - - - - - - -

exponents

3. The second parts of the mathematical expressions are used to find the mantissa in the log table. The mantissa for log 4 is the same in all of the following mathematical expressions:

$\log 10^1 + \log 4 = 1.6021$

$\log 10^2 + \log 4 = 2.6021$

$\log 10^3 + \log 4 =$ _____

$\log 10^4 + \log 4 =$ _____

- - - - - - - - - - - - - - - - - - -

3.6021

4.6021

4. In the mathematical expressions in frame 3 the characteristic is ______________ (different, the same) but the mantissa ____________ (is, is not) common to all.

- - - - - - - - - - - - - - - - - - -

different

is

5. Fill in the characteristics below:

From	To but Not Including	Characteristic
10,000	100,000	_____
1,000	10,000	_____
100	1,000	_____
10	100	_____
1	10	_____

4

3

2

1

0

6. The first step in finding the log of a number is to determine its characteristic. The characteristic of

log 76.5 is 1;

log 765 is 2;

log 7650 is ____;

log 76500 is ____;

log 765000 is ____.

- - - - - - - - - - - - - - - - - -

3

4

5

7. Note that the characteristic of a log is identical with the power designation (exponent) in the expression, log 10^n + log of a number greater than 1, but less than 10. Below is a comparison of both forms:

log 76.5	log 10^1	+ log 7.65
log 765	_____	+ log 7.65
log 7650	_____	+ log 7.65

log 76500 ______ + log 7.65

log 765000 ______ + log 7.65

- - - - - - - - - - - - - - - - - -

10^2
10^3
10^4
10^5

8. Determine the characteristics for the numbers below:

log 76.5 = log 10^1 + log 7.65 ______

log 76500 = log 10^4 + log 7.65 ______

log 765 = log 10^2 + log 7.65 ______

log 765000 = log 10^5 + log 7.65 ______

log 7650 = log 10^3 + log 7.65 ______

- - - - - - - - - - - - - - - - - -

1
4
2
5
3

9. The characteristic of log 577 is ______.

- - - - - - - - - - - - - - - - - -

2

10. The characteristic of log 35 is ______.

- - - - - - - - - - - - - - - - - -

1

11. The characteristic of log 8500 is ______.

- - - - - - - - - - - - - - - - - -

3

12. The characteristic of log 953, 500 is _____.

- - - - - - - - - - - - - - - - - -

5

13. The characteristic of log 955, 356 is ______.

- - - - - - - - - - - - - - - - - -

5

14. It is advisable at this point to write down the equivalent power of 10 and then determine the characteristic. After some practice the intermediate steps can be eliminated.

					Characteristic
log 577	=	log 10^2	+	log 5.77	______
log 953, 500	=	______	+	log 9.535	______
log 8, 500	=	______	+	log 8.50	______
log 955, 356	=	______	+	log 9.55356	______

log 953, 000 = ______ + log 9.53 ______

- - - - - - - - - - - - - - - - - -

	2
10^5	5
10^3	3
10^5	5
10^5	5

15. log 4000 = log 10^3 + log 4 = 3.6021

The characteristic 3 of the above log is the power designation in 10^3. The mantissa .6021 was found in the log table by looking up the number 4. When looking up the mantissa, the decimal point is disregarded. The numbers 43, 430, 4300, etc., have the same mantissa; the characteristic for each of the aforementioned numbers will be different, however,

Determine the mantissa for the numbers below:

log 577 .7612

log 5770 _____

log 57.7 _____

log 5.77 _____

- - - - - - - - - - - - - - - - - -

.7612

.7612

.7612

16. Determine the mantissa for the numbers below:

log 8970 .9528

log 897 _____

log 89700 _____

log 8.97 _____

- - - - - - - - - - - - - - - - - -

.9528

.9528

.9528

17. The log of a number consists of two parts, the __________ and the ____________.

- - - - - - - - - - - - - - - - - -

characteristic

mantissa

18. To obtain the log of a number convert it into the mathematical expression, of which the following is an example:

log 573 = log _____ + log _____

- - - - - - - - - - - - - - - - - -

10^2

5.73

19. The mantissa represents the second part in the above expression and is found in the log table. In looking up the mantissa, the decimal point of the number is disregarded. The mantissa, however, is always a(n) ____________ (integer, fraction).

- - - - - - - - - - - - - - - - - - - -

fraction

SELF-TEST

Break the numbers down as shown in the following example:

$$\log 765 = \log 10^2 + \log 7.65$$

1. log 8195 = ________________
2. log 217.8 = ________________
3. log 31,540 = ________________
4. log 1.753 = ________________
5. log 17.31 = ________________

Determine the characteristics of the following numbers:

		Characteristic
6.	log 188.9	____________
7.	log 1889	____________
8.	log 1.889	____________
9.	log 588,300	____________
10.	log 58.83	____________

Answers

1.	$\log 10^3 + \log 8.195$	6.	2
2.	$\log 10^2 + \log 2.178$	7.	3
3.	$\log 10^4 + \log 3.154$	8.	0
4.	$\log 10^0 + \log 1.753$	9.	5
5.	$\log 10^1 + \log 1.731$	10.	1

UNIT SIX

How To Use an Expanded Logarithm Table

1. The log table as used until now gives us mantissas for numbers with two significant digits or figures.

$6700 = 1000 \times 6.70 = 10^3 \times 6.70$

In log form

$\log 6700 = \log 1000 + \log 6.70 = \log 10^3 + \log 6.70 =$ ______

- - - - - - - - - - - - - - - - - - -

3.8261

2. If we are to find log 6740, we cannot use Figure 1; for instance,

$\log 6740 = \log 1000 \times 6.74 = \log 10^3 + \log 6.74$

But log 6.74 is not found in Figure 1. To find log 6.74 or any number with three significant digits we must use the table found in Figure 2. First we find the number 67 under the N column. Then we read to the right on the top line until we find the number 4. Now we drop down under the number 4 until we find the place where 67 and 4 intersect. The number found there is the mantissa of log 6.74.

$\log 6740 = 3.8287$

Common logarithms

N	0	1	2	3	4	5	6	7	8	9
10	0000	0043	0086	0128	0170	0212	0253	0294	0334	0374
11	0414	0453	0492	0531	0569	0607	0645	0682	0719	0755
12	0792	0828	0864	0899	0934	0969	1004	1038	1072	1106
13	1139	1173	1206	1239	1271	1303	1335	1367	1399	1430
14	1461	1492	1523	1553	1584	1614	1644	1673	1703	1732
15	1761	1790	1818	1847	1875	1903	1931	1959	1987	2014
16	2041	2068	2095	2122	2148	2175	2201	2227	2253	2279
17	2304	2330	2355	2380	2405	2430	2455	2480	2504	2529
18	2553	2577	2601	2625	2648	2672	2695	2718	2742	2765
19	2788	2810	2833	2856	2878	2900	2923	2945	2967	2989
20	3010	3032	3054	3075	3096	3118	3139	3160	3181	3201
21	3222	3243	3263	3284	3304	3324	3345	3365	3385	3404
22	3424	3444	3464	3483	3502	3522	3541	3560	3579	3598
23	3617	3636	3655	3674	3692	3711	3729	3747	3766	3784
24	3802	3820	3838	3856	3874	3892	3909	3927	3945	3962
25	3979	3997	4014	4031	4048	4065	4082	4099	4116	4133
26	4150	4166	4183	4200	4216	4232	4249	4265	4281	4298
27	4314	4330	4346	4362	4378	4393	4409	4425	4440	4456
28	4472	4487	4502	4518	4533	4548	4564	4579	4594	4609
29	4624	4639	4654	4669	4683	4698	4713	4728	4742	4757
30	4771	4786	4800	4814	4829	4843	4857	4871	4886	4900
31	4914	4928	4942	4955	4969	4983	4997	5011	5024	5038
32	5051	5065	5079	5092	5105	5119	5132	5145	5159	5172
33	5185	5198	5211	5224	5237	5250	5263	5276	5289	5302
34	5315	5328	5340	5353	5366	5378	5391	5403	5416	5428
35	5441	5453	5465	5478	5490	5502	5514	5527	5539	5551
36	5563	5575	5587	5599	5611	5623	5635	5647	5658	5670
37	5682	5694	5705	5717	5729	5740	5752	5763	5775	5786
38	5798	5809	5821	5832	5843	5855	5866	5877	5888	5899
39	5911	5922	5933	5944	5955	5966	5977	5988	5999	6010
40	6021	6031	6042	6053	6064	6075	6085	6096	6107	6117
41	6128	6138	6149	6160	6170	6180	6191	6201	6212	6222
42	6232	6243	6253	6263	6274	6284	6294	6304	6314	6325
43	6335	6345	6355	6365	6375	6385	6395	6405	6415	6425
44	6435	6444	6454	6464	6474	6484	6493	6503	6513	6522
45	6532	6542	6551	6561	6571	6580	6590	6599	6609	6618
46	6628	6637	6646	6656	6665	6675	6684	6693	6702	6712
47	6721	6730	6739	6749	6758	6767	6776	6785	6794	6803
48	6812	6821	6830	6839	6848	6857	6866	6875	6884	6893
49	6902	6911	6920	6928	6937	6946	6955	6964	6972	6981
50	6990	6998	7007	7016	7024	7033	7042	7050	7059	7067
51	7076	7084	7093	7101	7110	7118	7126	7135	7143	7152
52	7160	7168	7177	7185	7193	7202	7210	7218	7226	7235
53	7243	7251	7259	7267	7275	7284	7292	7300	7308	7316
54	7324	7332	7340	7348	7356	7364	7372	7380	7388	7396

N	0	1	2	3	4	5	6	7	8	9
55	7404	7412	7419	7427	7435	7443	7451	7459	7466	7474
56	7482	7490	7497	7505	7513	7520	7528	7536	7543	7551
57	7559	7566	7574	7582	7589	7597	7604	7612	7619	7627
58	7634	7642	7649	7657	7664	7672	7679	7686	7694	7701
59	7709	7716	7723	7731	7738	7745	7752	7760	7767	7774
60	7782	7789	7796	7803	7810	7818	7825	7832	7839	7846
61	7853	7860	7868	7875	7882	7889	7896	7903	7910	7917
62	7924	7931	7938	7945	7952	7959	7966	7973	7980	7987
63	7993	8000	8007	8014	8021	8028	8035	8041	8048	8055
64	8062	8069	8075	8082	8089	8096	8102	8109	8116	8122
65	8129	8136	8142	8149	8156	8162	8169	8176	8182	8189
66	8195	8202	8209	8215	8222	8228	8235	8241	8248	8254
67	8261	8267	8274	8280	8287	8293	8299	8306	8312	8319
68	8325	8331	8338	8344	8351	8357	8363	8370	8376	8382
69	8388	8395	8401	8407	8414	8420	8426	8432	8439	8445
70	8451	8457	8463	8470	8476	8482	8488	8494	8500	8506
71	8513	8519	8525	8531	8537	8543	8549	8555	8561	8567
72	8573	8579	8585	8591	8597	8603	8609	8615	8621	8627
73	8633	8639	8645	8651	8657	8663	8669	8675	8681	8686
74	8692	8698	8704	8710	8716	8722	8727	8733	8739	8745
75	8751	8756	8762	8768	8774	8779	8785	8791	8797	8802
76	8808	8814	8820	8825	8831	8837	8842	8848	8854	8859
77	8865	8871	8876	8882	8887	8893	8899	8904	8910	8915
78	8921	8927	8932	8938	8943	8949	8954	8960	8965	8971
79	8976	8982	8987	8993	8998	9004	9009	9015	9020	9025
80	9031	9036	9042	9047	9053	9058	9063	9069	9074	9079
81	9085	9090	9096	9101	9106	9112	9117	9122	9128	9133
82	9138	9143	9149	9154	9159	9165	9170	9175	9180	9186
83	9191	9196	9201	9206	9212	9217	9222	9227	9232	9238
84	9243	9248	9253	9258	9263	9269	9274	9279	9284	9289
85	9294	9299	9304	9309	9315	9320	9325	9330	9335	9340
86	9345	9350	9355	9360	9365	9370	9375	9380	9385	9390
87	9395	9400	9405	9410	9415	9420	9425	9430	9435	9440
88	9445	9450	9455	9460	9465	9469	9474	9479	9484	9489
89	9494	9499	9504	9509	9513	9518	9523	9528	9533	9538
90	9542	9547	9552	9557	9562	9566	9571	9576	9581	9586
91	9590	9595	9600	9605	9609	9614	9619	9624	9628	9633
92	9638	9643	9647	9652	9657	9661	9666	9671	9675	9680
93	9685	9689	9694	9699	9703	9708	9713	9717	9722	9727
94	9731	9736	9741	9745	9750	9754	9759	9763	9768	9773
95	9777	9782	9786	9791	9795	9800	9805	9809	9814	9818
96	9823	9827	9832	9836	9841	9845	9850	9854	9859	9863
97	9868	9872	9877	9881	9886	9890	9894	9899	9903	9908
98	9912	9917	9921	9926	9930	9934	9939	9943	9948	9952
99	9956	9961	9965	9969	9974	9978	9983	9987	9991	9996

Figure 2

The mantissa for log 6.75 is found by looking for the place at which row _____ and column _____ intersect.

- - - - - - - - - - - - - - - - - - -

67

5

3. The mantissa for log 6.75 is __________.

- - - - - - - - - - - - - - - - - - -

.8293

4. The mantissa for log 6.75 is __________.

- - - - - - - - - - - - - - - - - - -

.8299

5. The mantissa for log 1.01 is __________.

- - - - - - - - - - - - - - - - - - -

.0043

6. The mantissa for log 1.02 is __________.

- - - - - - - - - - - - - - - - - - -

.0086

7. Find the mantissas for the logs of the following numbers:

2.50 _______________

2.51 _______________

2.52 _______________

2.59 _______________

.3937

.3997

.4014

.4133

8. Find the mantissas for the logs of the following numbers:

9.00 ______________

9.01 ______________

9.02 ______________

9.50 ______________

9.51 ______________

9.99 ______________

- - - - - - - - - - - - - - - - - -

.9542

.9547

.9552

.9777

.9782

.9996

9. Find the mantissas for the logs of the following numbers:

4.00 ______________

4.36 ______________

6.52 ______________

1.59 ______________

1.14 ______________

7.34 ____________

6.23 ____________

6.00 ____________

8.49 ____________

9.35 ____________

- - - - - - - - - - - - - - - - - - - -

.6021

.6395

.8142

.2014

.0569

.8657

.7945

.7782

.9289

.9708

10. To sum up, the first column denotes numbers with two significant figures. To find the third significant figure look for the appropriate column on top and follow down to the spot at which it intersects the horizontal row. The point of intersection will show the mantissa of a log of a number with three significant figures. Often we are required to find the mantissa of a log of a number with four significant figures, in which event another step is required. This is called interpolation.

The mantissa for the log of the number 4.143 does not appear in Figure 6.2. The mantissas for log 4.14 and log 4.15 do appear, however. Find the mantissas for the logs of these numbers.

4.14 ____________

4.15 ____________

- - - - - - - - - - - - - - - - - - -

.6170

.6180

11. It is only reasonable to assume that the mantissa for log 4.143, which is not listed in the table, must fall between the mantissa of log 4.14 (which may also be written 4.140) and the mantissa of log 4.15 (which may also be written 4.150). Let's take a look at the following instruction:

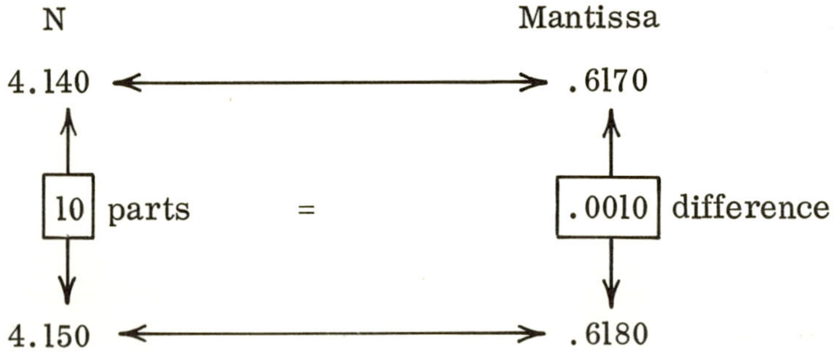

12. Let's expand our improvised log table.

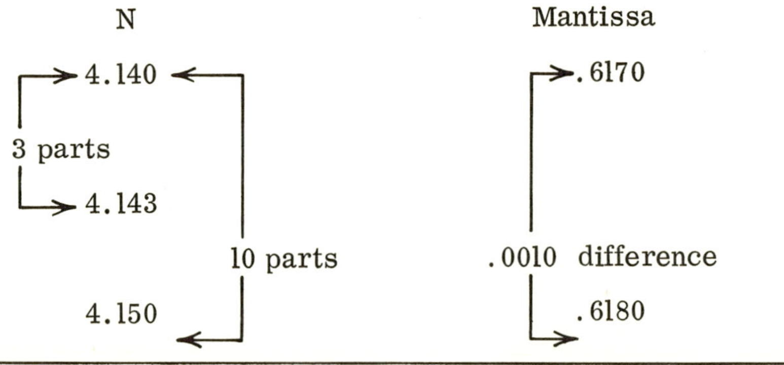

If 10 parts = .0010, then 1 part = $\frac{.0010}{10}$ = .0001.

Therefore 3 parts = .0001 x 3 = ____________.

- - - - - - - - - - - - - - - - - - -

.0003

13. The mantissa of log 4.143 = (mantissa of log 4.140 + 3 parts)
= (.6170 + .0003)
= ____________.

- - - - - - - - - - - - - - - - - - -

.6173

14. The mantissa of log 4.145 = (mantissa of log 4.140 + 5 parts
= (.6170 + .0005)
= ____________.

- - - - - - - - - - - - - - - - - - -

.6175

15. The mantissa of log 4.147 = (mantissa of log 1.140 + _____ parts)
= (.6170 + _____)
= ____________.

- - - - - - - - - - - - - - - - - - -

7

.0007

.6177

16. The mantissa of log 4.148 = ____________.

- - - - - - - - - - - - - - - - - -

.6178

17. The mantissa of log 4.141 = ____________.

- - - - - - - - - - - - - - - - - -

.6171

18. Let's find the mantissa for log 3.556. Since this number is not listed, we take the two numbers that are closest. On the low side is 3.55 (or 3.550) and on the high side, 3.56 (or 3.560). Again we improve a simple log table as shown below. Examine the table and insert the missing figures.

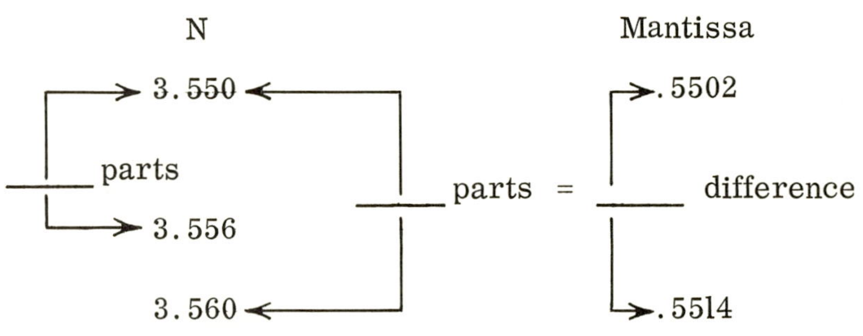

- - - - - - - - - - - - - - - - - -

6

10

.0012

19. If 10 parts = .0012, then 1 part = $\frac{.0012}{10}$ = .00012.

Therefore 6 parts = .00012 x 6 = ____________.

- - - - - - - - - - - - - - - - - -

.0007 (rounded off to 4 decimal places)

20. The mantissa of log 3.556 = (mantissa of log 3.550 + 6 parts)

= (.5502 + .0007)

= ____________.

- - - - - - - - - - - - - - - - -

.5509

21. The mantissa of log 3.558 = (mantissa of log 3.550 + _____ parts)

= (.5502 + _____)

= ____________.

- - - - - - - - - - - - - - - - -

8

.0010

.5512

22. The mantissa of log 3.559 = (mantissa of log 3.550 + ____ parts)

= (.5502 + _____)

= ____________.

- - - - - - - - - - - - - - - - -

9

.0011

.5513

23. Find the mantissas for the logs of the following numbers:

(a) 3.551 ____________

(b) 3.552 ____________

(c) 3.553 ____________

(d) 3.554 ____________

(e) 3.555 ____________

(f) 3.557 ____________

- - - - - - - - - - - - - - - - - - -

(a) .5503

(b) .5504

(c) .5506

(d) .5507

(e) .5508

(f) .5510

Note: After some practice you will find that it isn't necessary to draw up a table to find an unlisted mantissa, but it helps if you keep a mental picture when working with them.

24. Let's try a shortcut in finding the mantissa for the log of of the number 3.146.

Step 1 The listed numbers below and above 3.146 are

mantissa of log 3.150	=	.4983
mantissa of log 3.140	=	.4969
10 parts	=	.0014

Step 2 If 10 parts = .0014, then 1 part = $\frac{.0014}{10}$ = .00014

Step 3 Mantissa of log 3.146 = .4969 + .00014 x 6

= .4969 + ______

= __________.

- - - - - - - - - - - - - - - - - -

.0008

.4977

25. Find the mantissa of log 1.487.

Step 1 The listed numbers below and above 1.487 are

mantissa of log 1.490 = .1732

mantissa of log 1.480 = .1703

10 parts = .0029

Step 2 If 10 parts = .0029, then 1 part = $\frac{.0029}{10}$ = .00029.

Step 3 Mantissa of log 1.487 = .1703 + .00029 x 7

= .1703 + ______

= __________.

- - - - - - - - - - - - - - - - - -

.0020

.1723

26. Using the above method, find the mantissas for the logs of the following numbers:

2.653 ____________

2.938 ____________

5.789 ____________

8.888 ____________

- - - - - - - - - - - - - - - - - -

.4237

.4680

.7626

.9488

27. The mantissa of log 9.555 = ____________.

- - - - - - - - - - - - - - - - - -

.9803 (rounded off to 4 decimal places)

28. The mantissa of log 9.905 = ____________.

- - - - - - - - - - - - - - - - - -

.9959 (rounded off to 4 decimal places)

29. Some log tables show the parts already computed. Figure 3 is such a table. It makes finding mantissas for logs of four significant digits very easy; for instance, to find the mantissa for log 9.026 we have the following:

number given	9.026
closest number in table	9.020
difference	0.006

Now just think of 0.006 as 6 pp (proportional parts). Look at Figure 3 and you will see columns headed "Proportional Parts" with numbers 1 through 9. In our example we must find the number 6 under "Proportional Parts" and then go down the column to the point at which 90 intersects with number 6. This will give us the number 3. It is necessary to add decimal places to the number we have found. Therefore the number 3 becomes .0003. Now we are ready to find the mantissa of log 9.026:

mantissa of log	9.020	=	.9552
pp from log table	0.006	=	.0003
mantissa of log	9.026	=	.9555

30. Find the proportional parts of log 9.539:

number given	9.539
closest number in table	9.530
difference	_____

- - - - - - - - - - - - - - - - - -

0.009

31. 0.009 = _______ proportional parts

- - - - - - - - - - - - - - - - - -

9

32. Find the mantissa of log 9.539:

mantissa of log	9.530	=	.9791
pp from log table	0.009	=	_____
mantissa of log	9.539	=	_____

- - - - - - - - - - - - - - - - - -

.0004

.9795

33. Find the mantissa of log 8.639:

number given 8.639

closest number in table _____

difference _____

mantissa of log _____ = _____

pp from log table _____ = _____

mantissa of log 8.639 = _____

- - - - - - - - - - - - - - - - - -

8.630

0.009

8.630 = .9360

9 = .0005

8.639 = .9365

34. Mantissa of log 7.635 = ____________.

- - - - - - - - - - - - - - - - - -

.8828

35. Mantissa of log 7.945 = ____________.

- - - - - - - - - - - - - - - - - -

.9001

36. Mantissa of log 9.999 = ____________ .

- - - - - - - - - - - - - - - - - -

1.0000

37. From the above problem we found that the mantissa of log 9.999 = 1.0000. Would you say that this is 100% accurate? Of course not. However, 9.999 is close enough to 10 and the equivalent log or mantissa for all practical purposes will be 1.0000.

In summary, to find the log of a number change the number into its equivalent exponential form of 10^n x a number greater than 1 but less than ______.

- - - - - - - - - - - - - - - - - -

10

38. The raised figure "n" in "10^n x a" becomes the ____________ (characteristic, mantissa).

- - - - - - - - - - - - - - - - - -

characteristic

39. The "a" in "10^n x a" is used to find the __________________ (characteristic, mantissa) in the log table.

- - - - - - - - - - - - - - - - - -

mantissa

40. First comes the __________________ (characteristic, mantissa) and then the decimal point, followed by the _________________ (characteristic, mantissa).

- - - - - - - - - - - - - - - - - -

characteristic

mantissa

The mantissa of a number with three significant figures is found in three steps:

Step 1 Look up the first two significant digits under "N" in the log table.

Step 2 Follow across the "N" for the third significant digit.

Step 3 Move down the column to the row where you found the first two significant digits. The point of intersection will indicate the mantissa.

The mantissa of a number with four significant figures is found in the same way as one with three significant figures, EXCEPT that the value of the proportional parts is added.

SELF-TEST

1. log 3.57 = ____________

2. log 5.89 = ____________

3. log 6.73 = ____________

4. log 8.75 = ____________

5. log 9.78 = ____________

6. log 4.218 = ____________

7. log 2.635 = ____________

8. log 26.35 = ____________

9. log 163.70 = ____________

10. log 98.71 = ____________

11. log 956,500 = ____________

12. log 65, 380 = ____________

13. log 71.35 = ____________

14. log 415.30 = ____________

15. log 7,365 = ____________

16. log 12.65 = ____________

17. log 225.60 = ____________

18. log 17.56 = ____________

19. log 119.50 = ____________

20. log 9.514 = ____________

Answers

1. .5527
2. .7701
3. .8280
4. .9420
5. .9903
6. .6251
7. .4208

8. 1.4208

9. 2.2140

10. 1.9943

11. 5.9807

12. 4.8154

13. 1.8534

14. 2.6183

15. 3.8672

16. 1.1021

17. 2.3534

18. 1.2445

19. 2.0774

20. .9784

UNIT SEVEN

Negative Logarithms

1. Up to now we have been dealing with mantissa for numbers 1 to 10. What about a number that is below 1, or, in other words, a fraction? Let's examine the table that follows:

mantissa of log	9.99	=	.9996
	8.99	=	.9538
	6.99	=	.8445
	6.00	=	.7782
	2.00	=	.3010
	1.50	=	.1761
	1.00	=	.0000
	0.90	=	?

The mantissa of log 0.90 is below the mantissa of log 1.00. The mantissa of log 1.00 = .0000, or 0. In common mathematics any number below 0 is a negative number. Since logarithms are governed by the rules of mathematics, log 0.90 must also be a ________________ number.

- - - - - - - - - - - - - - - - - - -

negative

2. Log .8536 is a ________________ (positive, negative) quantity.

- - - - - - - - - - - - - - - - - - -

negative

3. Log 1.547 is a ________________ (positive, negative) quantity.

- - - - - - - - - - - - - - - - - - -

positive

4. Log .1777 is a ________________ (positive, negative) quantity.

- - - - - - - - - - - - - - - - - - -

negative

5. The log of a fraction is a ________________ (positive, negative) quantity.

- - - - - - - - - - - - - - - - - - -

negative

6. We know that the log of a number less than 1 must be a negative quantity. Our log table, however, does not show numbers that are less than 1, nor does it show negative mantissas. Does this mean that we cannot deal in negative quantities? Not at all. If we just recall three important things that we have already learned, we will be able to deal with negative quantities.

   1. Logs are a type of exponential number.

   2. When using logs or exponential numbers, multiplications become simple additions; divisions become subtractions.

3. A decimal fraction may be written as a common fraction; for instance,

$.5 = \frac{5}{10}$ $.05 = \frac{5}{100}$ $.005 = \frac{5}{\_\_\_\_}$

$.1 = \_\_\_\_$ $.03 = \_\_\_\_$ $.003 = \_\_\_\_$

$\frac{5}{1000}$

$\frac{1}{10}$

$\frac{3}{100}$

$\frac{3}{1000}$

7. We must take one more step for numbers such as .12 or .198. The decimal fraction .12 could be written

$$\frac{12}{100}$$

but for our purpose it is better to write it

$$\frac{1.2}{10}$$

We are able to write it $\frac{1.2}{10}$ because we divided the numerator and the denominator by 10, or a multiple of 10, to convert the numerator to a number between 1 and 10; for example,

$$.83 = \frac{83}{100} \div \frac{10}{10} = \frac{8.3}{10} \qquad .156 = \frac{156}{1000} \div \frac{100}{100} = \frac{1.56}{10}$$

8. $.245 = \_\_\_\_ \div \_\_\_\_ = \frac{\_\_\_\_}{10}$

- - - - - - - - - - - - - - - - -

$\frac{245}{1000} \quad \frac{100}{100} = \frac{2.45}{10}$

9. $.58 = \_\_\_\_ \div \_\_\_\_ = \_\_\_\_$

- - - - - - - - - - - - - - - - -

$\frac{58}{100} \quad \frac{10}{10} = \frac{5.8}{10}$

10. The decimal fraction .5 may be written as a common fraction $\frac{5}{10}$. Any common fraction may be considered as a division; the numerator is to be divided by the denominator. Thus we can show $\frac{5}{10}$ as $5 \div 10$. Further, as we have already learned, division in log form is reduced to simple subtraction. Therefore $5 \div 10$ is written as

log 5 - log 10 (log 5 minus log 10)

$\log \frac{6}{10} = \log 6 - \log \_\_\_\_$

- - - - - - - - - - - - - - - - -

10

11. $\log .9 = \log \frac{9}{10} = \log \_\_\_\_ - \log \_\_\_\_$

- - - - - - - - - - - - - - - - -

9

10

12. $\log .25 = \log \frac{2.5}{10} = \_\_\_\_ - \_\_\_\_$

- - - - - - - - - - - - - - - - -

log 2.5 - log 10

13. $\log .8 = \log \frac{8}{10} = \log 8 - \log 10$. Now refer to your log table:

log 8 = __________ and, log 10 = __________

- - - - - - - - - - - - - - - - - - - -

.9031

1

14. $\log .02 = \log \frac{2}{100} = \log 2 - \log 100 = .3010 - 2$

$\log .03 = \log \frac{3}{100} = \log 3 - \log 100 =$ _____ - _____

- - - - - - - - - - - - - - - - - - - -

.4771 - 2

15. $\log .05 = \log \frac{5}{100} = \log 5 - \log 100 =$ _______ - 2

- - - - - - - - - - - - - - - - - - - -

.6990

16. $\log .06 = \log \frac{6}{100} = \log 6 - \log 100 =$ _______ - ____

- - - - - - - - - - - - - - - - - - - -

.7782 - 2

17. log .08 = log _____ = log _____ - log ____ = ______ - ______

- - - - - - - - - - - - - - - - - - - -

$\frac{8}{100}$

8

100

.9031 - 2

18. $\log .003 = \log \frac{3}{1000} = \log 3 - \log 1000 = .4771 - 3$

$\log .005 = \log \frac{5}{1000} = \log 5 - \log 1000 =$ ______ - ____

- - - - - - - - - - - - - - - - - - -

.6690 - 3

19. $\log .0067 = \log \frac{6.7}{1000} = \log 6.7 - \log 1000 =$ ________ - ______

- - - - - - - - - - - - - - - - - - -

.8261 - 3

20. $\log .006784 = \log \frac{6.784}{1000} = \log 6.784 - \log 1000$

$= .8315 -$ ______

- - - - - - - - - - - - - - - - - - -

3

21. $\log .000678 = \log \frac{6.78}{10,000} = \log 6.78 - \log 10,000$

$= .8312 -$ _____

- - - - - - - - - - - - - - - - - - -

4

22. $\log .0006784 = \log \dfrac{6.784}{10,000} = \log$ _____ $- \log$ _____

$= .8315 -$ _____

- - - - - - - - - - - - - - - - - -

6.784

10,000

.8315

4

23. $\log .1 = \log$ _____ $= \log$ _____ $- \log$ _____ $=$ _____

$-$ _____

- - - - - - - - - - - - - - - - - -

$\dfrac{1}{10}$

1

10

0 - 1

24. From the above example we can easily see that 0 - 1 = -1. Now we can proceed to the different ways of writing the logs of fractions. One way is to place the negative characteristic in front of the positive mantissa; for example, we could write

$$\log .05 = \log \frac{5}{100} = \log 5 - \log 100 = .6990 - 2$$

If, however, we reverse the positions of the characteristic and the mantissa, our example would then look like this:

-2 + .6990

Whichever way we use, we do not complete the subtraction operation but leave the characteristic and mantissa as separate entities. Another example would be

$\log .073 = \log \frac{7.3}{100} = -2 + .8633$

$\log .005 = \log \frac{5}{1000} =$ _____ + _________

- - - - - - - - - - - - - - - - - -

-3 + .6990

25. Another way to write the log of fractions is to place a minus sign above the negative characteristic.

$\bar{3}.6990$

By doing this we show that the characteristic is negative. Remember, however that the mantissa is not affected by this action. The mantissa remains positive.

A minus sign above the characteristic means that the characteristic is _______________ (positive, negative).

- - - - - - - - - - - - - - - - - -

negative

26. We have shown three ways to write negative logs:

1. $\log .03 = \log \frac{3}{100} = \log 3 - \log 100 = .4771 - 2$
2. $\log .03 = \log \frac{3}{100} = \log 3 - \log 100 = -2 + .4771$
3. $\log .03 = \log \frac{3}{100} = \log 3 - \log 100 = \bar{2}.4771$

Write the negative log .07 in the three ways used above.

1. ____________

2. ____________

3. ____________

- - - - - - - - - - - - - - - - - - -

1. .8451 - 2

2. -2 + .8451

3. $\bar{2}.8451$

Note: Other ways of expressing negative logs exist, but we regard the knowledge of the above as adequate and easy to follow.

27. The best way to write negative logs is to place a minus sign above the negative characteristic. So let's practice.

$\log .085 = \log \frac{8.5}{100} = \bar{2}.9294$

log .0678 = log ________ = ________

- - - - - - - - - - - - - - - - - - -

$\frac{6.78}{100} = \bar{2}.8312$

28. log .0006784 = log __________ = __________

- - - - - - - - - - - - - - - - - - -

$\frac{6.784}{10,000} = \bar{4}.8315$

29. log .371 = ____________ = ____________

- - - - - - - - - - - - - - - - - - -

$\frac{3.71}{10} = \bar{1}.5694$

30. log .566 = ____________ = ____________

- - - - - - - - - - - - - - - - - - -

$\frac{5.66}{10} = \bar{1}.7528$

31. log .00589 = ____________ = ____________

- - - - - - - - - - - - - - - - - - -

$\frac{5.89}{1000} = \bar{3}.7701$

32. The log of a fraction is a ________________ (positive, negative) quantity.

- - - - - - - - - - - - - - - - - - -

negative

33. Generally, log tables ____________ (do, do not) show negative mantissas.

- - - - - - - - - - - - - - - - - - -

do not

34. There are a number of ways to write negative logs, but in each case the characteristic will be ______________ (positive, negative) and the mantissa will be ______________ (positive, negative).

- - - - - - - - - - - - - - - - - - -

negative

positive

SELF-TEST

Write the logs of the numbers below in the following format:

$$\log .07554 = \log \frac{7.554}{100} = \log 7.554 - \log 100$$

$$= .8781 - 2 = -2 + \overline{2}.8781$$

1. .7981

2. .01234

3. .0005789

4. .008755

5. .00005864

Find the logs of the numbers below and write them in the following form: $\log .07554 = \overline{2}.8781$. Try to bypass the intermediate steps.

6. .5694

7. .008718

8. .07431

9. .00001753

10. .05819

Answers

1. $\log .7981 = \log \frac{7.981}{10} = \log 7.981 - \log 10 = .9021 - 1$

$= -1 + .9021 = \bar{1}.9021$

2. $\log .01234 = \log \frac{1.234}{100} = \log 1.234 - \log 100 = .0913$

$- 2 = -2 + .0913 = \bar{2}.0913$

3. $\log .0005789 = \log \frac{5.789}{10,000} = \log 5.789 - \log 10,000$

$= .7626 - 4 = -4 + .7626 = \bar{4}.7626$

4. $\log .008755 = \log \frac{8.755}{1000} = \log 8.755 - \log 1000$

$= .9422 - 3 = -3 + .9422 = \bar{3}.9422$

5. $\log .00005864 = \log \frac{5.864}{100,000} = \log 5.864 - \log 100,000$

$= .7682 - 5 = -5 + .7682 = \bar{5}.7682$

6. $\log .5694 = \bar{1}.7554$

7. $\log .008718 = \bar{3}.9404$

8. $\log .07431 = \bar{2}.8711$

9. $\log .00001753 = \bar{5}.2437$

10. $\log .05819 = \bar{2}.7649$

UNIT EIGHT

Finding the Antilog of Mantissas

1. Given a common number, we now know how to find the log. The characteristic is the exponent of an exponential number with the base of 10 and the mantissa can be looked up in the log table. Now, if we have the log and want to find the number, we just reverse the process. This is called finding the ANTILOG.

For the time being we will cover only antilogs with characteristics that are zero; for instance, let's assume we were given the mantissa .6021. To find the antilog of the mantissa we merely find that figure in the log table and then look to the left to find the first significant number. Then we look to the top of the column in which .6021 is found to determine the second significant number. In our example we find that the mantissa .6021 is located in row 40 and column 0. Therefore our number is 4.00.

Find the antilogs of these mantissas.

.6990 ________ .7782 ________ .9031 ________

- - - - - - - - - - - - - - - - - - -

5.00

6.00

8.00

2. To find the antilog of .9036 we look in the log table and find that .9036 is listed in row 80 and column 1. Therefore the antilog is 8.01

Find the antilogs of these mantissas.

.9042 ________ .9053 ________ .0212 ________

- - - - - - - - - - - - - - - - - -

8.02

8.04

1.05

3. Suppose you were asked to find the antilog for a number that does not appear in the log table; for instance, our log table does not contain the mantissa .8764. It does, however, show the mantissas .8762 and .8768. It is only reasonable to assume that antilog .8764, which is not listed, must fall between antilog .8762 and antilog .8768. (You have probably noticed that "of" is left out in antilog .8764, etc.) Let's take a look at the illustration.

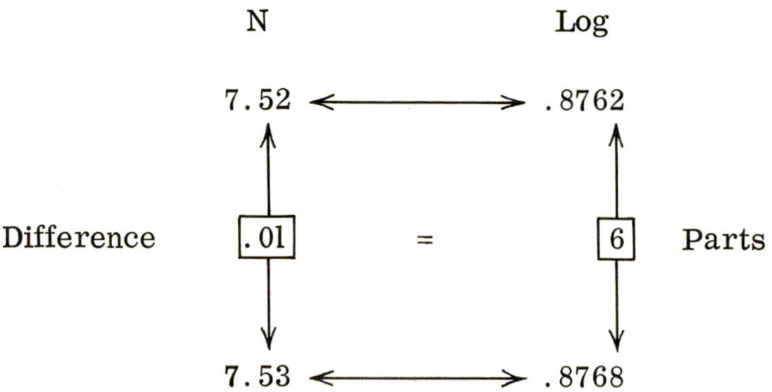

Let's expand our improvised log table.

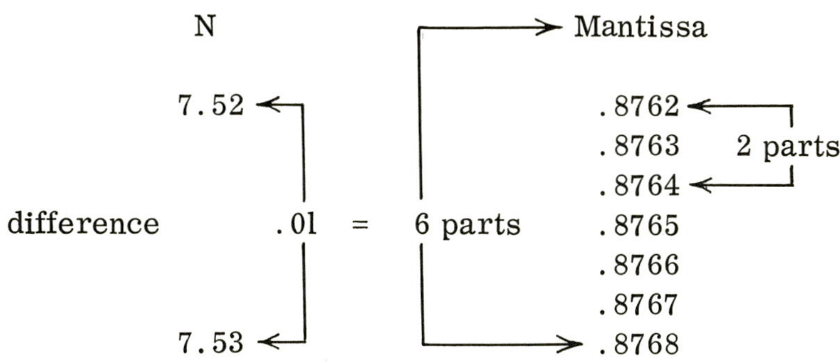

If 6 parts = .01, then 1 part $= \frac{.01}{6}$ and 2 parts $= \frac{.01}{6} \times 2$

= ________.

- - - - - - - - - - - - - - - - - -

.0033

4. antilog .8764 = (antilog .8762 + 2 parts)

= (7.52 + .0033)

= ____________

- - - - - - - - - - - - - - - - - -

7.5233

5. antilog .8767 = (antilog .8762 + 5 parts)

$= (7.52 + \frac{.01}{6} \times$ ______)

= (7.52 + __________)

= ______________

- - - - - - - - - - - - - - - - - -

5

.0083

7.5283

6. To find antilog .4949 we proceed the same way. This log is not listed in the log table but the closest logs are. On the low side is .4942 and on the high side, .4955. Insert the missing figures in the blanks provided in the illustration.

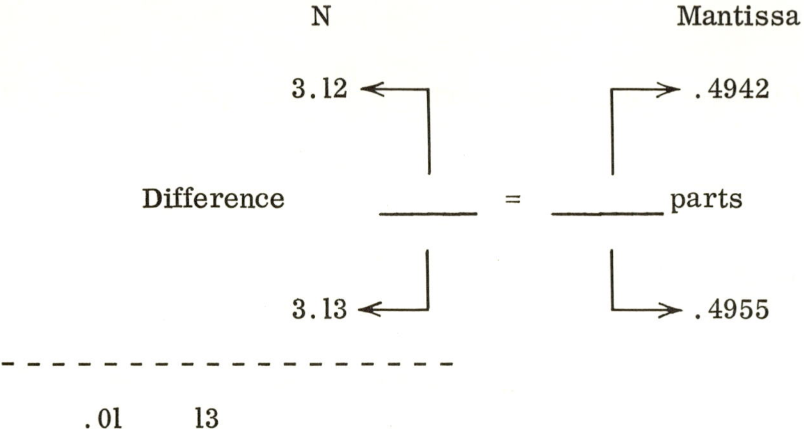

- - - - - - - - - - - - - - - - - -

.01 13

7. If 13 parts = .01, then 1 part = $\frac{.01}{\quad}$ and 7 parts $\frac{.01}{\quad}$ x ______.

- - - - - - - - - - - - - - - - - -

$\frac{.01}{13}$ $\frac{.01}{13}$

7

8. antilog .4949 = (antilog .4942 + 7 parts)

$$= (3.12 + \frac{.01}{13} \times 7)$$

= (3.12 + ______)

= ____________

- - - - - - - - - - - - - - - - - -

.0054

3.1254

9. antilog .4945 = (antilog .4942 + 3 parts)

$$= \left(3.12 + \frac{.01}{13} \times 3\right)$$

= (3.12 + .0023)

= 3.1223

We now insert the calculated antilogs in the table shown below.

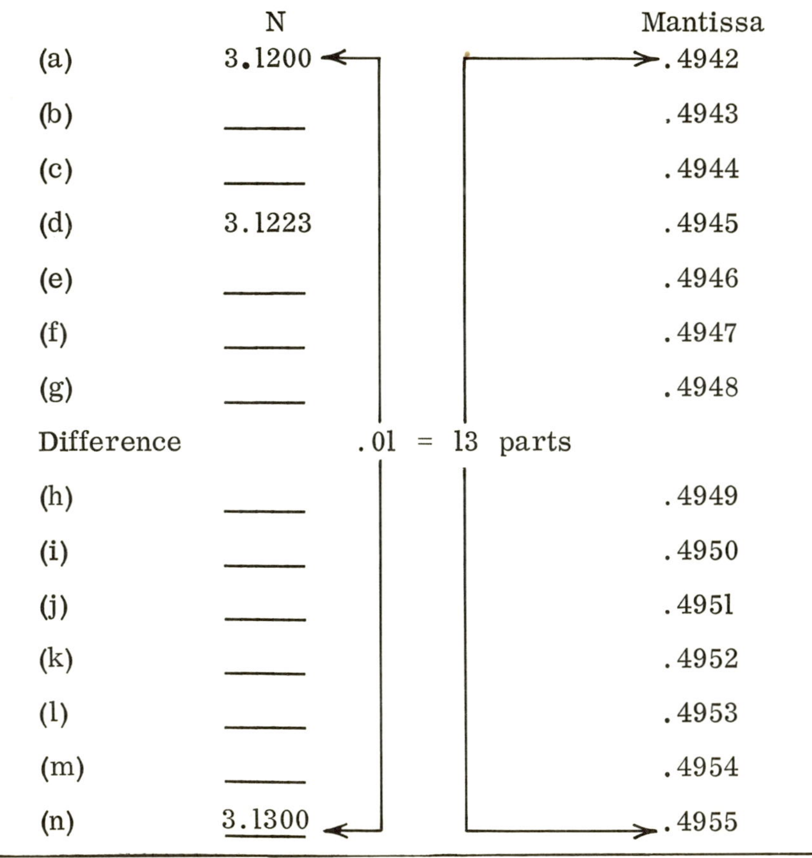

	N		Mantissa
(a)	3.1200		.4942
(b)	______		.4943
(c)	______		.4944
(d)	3.1223		.4945
(e)	______		.4946
(f)	______		.4947
(g)	______		.4948
Difference		.01 = 13 parts	
(h)	______		.4949
(i)	______		.4950
(j)	______		.4951
(k)	______		.4952
(l)	______		.4953
(m)	______		.4954
(n)	3.1300		.4955

Calculate all antilogs and insert in the table under the N column.

- - - - - - - - - - - - - - - - - -

(b) 3.1208

(c) 3.1215

(e) 3.1231

(f) 3.1239

(g) 3.1246

(h) 3.1254

(i) 3.1262

(j) 3.1269

(k) 3.1278

(l) 3.1285

(m) 3.1292

We have gone to great lengths to explain how to find antilogs that are not listed in the log table. It isn't necessary to draw up tables as we have done, but keep a mental picture of them when you are looking for these unlisted antilogs.

10. Find the antilog .0929.

Step 1 The listed mantissas below and above .0929 are

antilog .0934	=	1.24
antilog .0899	=	1.23
35 parts	=	.01

Step 2 If 35 parts = .01, then 1 part $= \frac{.01}{35} = .000285$

Step 3 antilog .0929 = (antilog .0899 + 30 parts)

= (1.23 + .000285 x 30)

= (1.23 + ________)

= ____________

- - - - - - - - - - - - - - - - - -

.0086

1.2386

11. Find the antilog .2435.

Step 1 The listed mantissas below and above .2435 are

antilog .2455	=	1.76
antilog .2430	=	1.75
25 parts	=	.01

Step 2 If 25 parts = .01, then 1 part = $\frac{.01}{}$ = ________

Step 3 antilog .2435 = (antilog .2430 + 5 parts)

= (1.75 + .0004 x 5)

= (1.75 + ______)

= ____________

- - - - - - - - - - - - - - - - - -

$\frac{.01}{25}$ = .0004

.002

1.752

12. Find the antilogs of these mantissas.

.3101 ____________

.4370 ____________

.5139 ____________

.5603 ____________

- - - - - - - - - - - - - - - - - -

2.042

2.735

3.265

3.633

13. Now that we know how to find antilogs that do not appear in the log table by using the above method let's try a simpler way. This time we will use proportional parts in the log table. Finding antilogs by using proportional parts is just the reverse of finding proportional parts to logs; for example, let's find the antilog for .3981. Although .3981 does not appear in the log table, the next lower mantissa can be found. It is ____________.

- - - - - - - - - - - - - - - - - -

.3979

14. The first step in finding the antilog for a mantissa that does not appear in the log table by using proportional parts is to find the next ______________ (lower, higher) mantissa that does appear.

- - - - - - - - - - - - - - - - - -

lower

15. The second step is to find the difference between the two:

given mantissa	.3981
next lower	______
difference	______

- - - - - - - - - - - - - - - - - -

.3979

.0002

16. The first step in finding the antilog for a mantissa that does not appear in the log table is to find the next ____________ (lower, higher) mantissa that does appear. The second step is to find the ____________ between the given and the next lower known mantissa.

- - - - - - - - - - - - - - - - - -

lower

difference

17. The third step is to consider the difference between the known and the unknown mantissas in terms of equivalent proportional parts. We do this by ignoring the decimal point and the zeros that immediately follow it. In our example the difference is .0002. Ignoring the decimal point and the three zeros that follow it, we think of this as "2" equivalent proportional parts.

Let's continue with the example from the last frame:

given mantissa	.3981
next lower	.3979
difference	.0002

How many equivalent proportional parts have we found?

- - - - - - - - - - - - - - - - - -

2

18. Note that we ignored only the zeros that immediately followed the decimal point and the decimal point itself.

.0002 = /ØØØ2 = 2

Now find the equivalent pp for this example:

given mantissa	.5549
next lower	.5539
difference	_____ = ______ pp

- - - - - - - - - - - - - - - - - -

.0010 = 10

19. The first step in determining the antilog of an unlisted mantissa is to find the ____________________ mantissa in the log table.

- - - - - - - - - - - - - - - - - -

next lower

20. The second step is to find the _________________ between the given and the next lower mantissa.

- - - - - - - - - - - - - - - - - -

difference

21. The third step is to find the equivalent pp by ignoring the ___________ immediately following the ____________ and the ____________ itself.

- - - - - - - - - - - - - - - - - -

zeros decimal

decimal

22. Returning to our first example, which was to find the antilog for .3981, we found that the equivalent pp was 2. Looking at the table, find the next lower mantissa to .3981 which is .3979. In that same row move to the right until you come to the number 2 under the column headed "Proportional Parts." Now go to the top of this column. What number do you find there? ________

- - - - - - - - - - - - - - - - - -

1

23. Now think of the number you found as a fraction, $\frac{1}{1000}$, or as a decimal .001. All we have to do is add .001 to antilog .3979:

antilog	.3979	=	2.50
pp	2	=	.001
antilog	.3981	=	______

- - - - - - - - - - - - - - - - - -

2.501

24. Why should you think of the number 1 as .001? Finding the antilog is the reverse of finding the log. Look for log 2.501 on our log table. You will find that 2.50 can be determined directly and the last digit 1 must be found as proportional part (pp). Notice the place of this digit. It occupies the third place <u>after</u> the decimal point and is therefore .001. It will become clearer when you go through both operations on paper.

Find log 2.501. ________

Find antilog .3981. ________

- - - - - - - - - - - - - - - - - -

.3981

2.501

25. PROBLEM: Find the antilog .4825.

.4825 __________ (does, does not) appear in the log table.

- - - - - - - - - - - - - - - - - -

does not

26. The next lower number that does appear in the log table is ________.

- - - - - - - - - - - - - - - - - -

.4814

27. antilog .4825

next lower .4814

difference _____ = _______ pp

- - - - - - - - - - - - - - - - - -

.0011 = .008

28. antilog .4814 = ______

pp .0011 = .008

antilog .4825 = ______

- - - - - - - - - - - - - - - - - -

3.030

3.038

29. antilog .2012 = ________

- - - - - - - - - - - - - - - - - -

antilog	.2012		
next lower	.1987		
difference	.0025	=	~~.00~~25 = 25 pp = $\frac{9}{1000}$ = .009
antilog	.1987	=	1.580
pp	.0025	=	.009
antilog	.2012	=	1.589

30. antilog .4430 = ____________

- - - - - - - - - - - - - - - - - - -

2.773

31. In finding unlisted antilog .2272, the next lower mantissa is .2253. The difference between the two is .0019, or 19 pp. Going to the right on the row listing .2253, we find no 19 pp. However, we do find 18 and 21 pp. Logically, we will select the closest pp, which is 18, and follow through with our problem. What then is the antilog .2272? ____________

- - - - - - - - - - - - - - - - - - -

1.687

32. antilog .1012 = ____________

- - - - - - - - - - - - - - - - - - -

1.262

In this problem you probably noticed that there was no equivalent proportional part for 8. The closest to it in the table is 7, and we trust that you chose this figure for your calculations.

33. antilog .9364 = ____________

- - - - - - - - - - - - - - - - - - -

8.637 or 8.638

In this problem both answers are correct, since 4 appears in two columns. Keep in mind that the last decimal digit is of a low order and any error due to inaccuracies is insignificant.

SELF-TEST

1. antilog .9805 = ____________
2. antilog .8675 = ____________
3. antilog .0755 = ____________
4. antilog .3534 = ____________
5. antilog .7986 = ____________
6. antilog .0199 = ____________
7. antilog .9975 = ____________
8. antilog .9277 = ____________
9. antilog .9265 = ____________
10. antilog .1551 = ____________
11. antilog .1330 = ____________
12. antilog .2940 = ____________
13. antilog .4991 = ____________

14. antilog .4380 = ____________

15. antilog .6301 = ____________

Answers

1. 9.56
2. 7.37
3. 1.19
4. 2.256
5. 6.288 or 6.289
6. 1.047
7. 9.942 or 9.943
8. 8.465 or 8.466
9. 8.443 or 8.444
10. 1.429
11. 1.358
12. 1.967 or 1.968
13. 3.156
14. 2.741
15. 4.267

UNIT NINE

Finding the Antilog of Characteristics and Mantissas

1. We have already learned that to multiply exponential numbers with the same base we merely add the exponents which is the same as adding logs. Conversely, if we convert from logs to common numbers, addition changes to ________

- - - - - - - - - - - - - - - - - -

multiplication

2. Antilog 2.4771 consists of antilog 2 and antilog ________

- - - - - - - - - - - - - - - - - -

.4771

3. Antilog 2 is simply 10^2 or ________.

- - - - - - - - - - - - - - - - - -

100

4. Using the log table, find the antilog .4771. ________

- - - - - - - - - - - - - - - - - -

3

5. As we said, addition of logs changes to multiplication if we convert to common numbers. Therefore to find the antilog 2.4771 we multiply antilog 2 by antilog .4771:

antilog 2 = ________

antilog .4771 = ________

- - - - - - - - - - - - - - - - - -

100

3

6. antilog 2.4771 = ________ x ________ = ________

- - - - - - - - - - - - - - - - - -

100 x 3 = 300

7. To find antilog 3.9777, the first step is to find the antilog ________.

- - - - - - - - - - - - - - - - - -

3

8. antilog 3 = $10^{\_\_\_}$ = ______

- - - - - - - - - - - - - - - - - -

10^3 = 1000

9. The second step is to find the antilog .9777. ________.

- - - - - - - - - - - - - - - - - -

9.5

10. The third and final step is to multiply the two antilogs:

antilog 3.9777 = ________ x ________ = ________

- - - - - - - - - - - - - - - - - -

1000 x 9.5 = 9500

11. PROBLEM: Find the antilog 4.6561:

The first step is to find the ____________ = ____________.

- - - - - - - - - - - - - - - - - - -

antilog 4 = 10,000

12. The second step is to find the ______________ = _________.

- - - - - - - - - - - - - - - - - - -

antilog .6561 = 4.53

13. The third step is to ______________ the two antilogs.

- - - - - - - - - - - - - - - - - - -

multiply

14. antilog 4.6561 = _________ x _________ = _________

- - - - - - - - - - - - - - - - - - -

10,000 x 4.53 = 45,300

15. PROBLEM: Find the antilog 3.7782:

antilog 3 = __________

antilog .7782 = __________

- - - - - - - - - - - - - - - - - - -

1000

6

16. antilog 3.7782 x ________ x ________ = ________

- - - - - - - - - - - - - - - - - -

1000 x 6 = 6000

17. PROBLEM: Find the antilog 1.7059:

antilog 1 = __________

antilog .7059 = __________

- - - - - - - - - - - - - - - - - -

10

5.08

18. antilog 1.7059 = ________ x ________ = ________

- - - - - - - - - - - - - - - - - -

10 x 5.08 = 50.8

19. Find the following antilogs:

antilog 3.6781 = __________

antilog 2.9327 = __________

antilog 4.0341 = __________

antilog 5.5279 = __________

- - - - - - - - - - - - - - - - - -

4,765

856.30 or 856.40

10,820

337,200

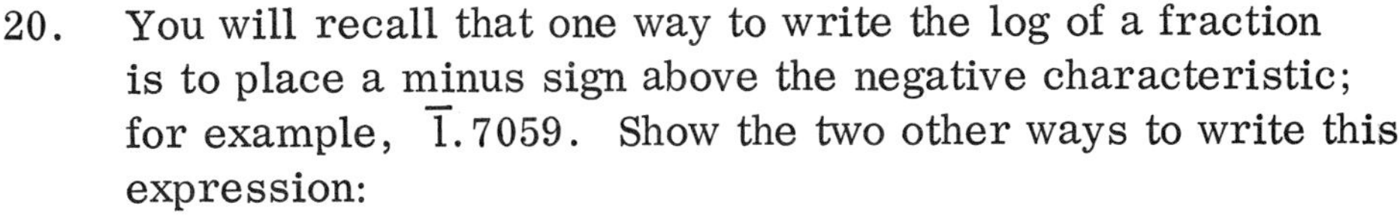

20. You will recall that one way to write the log of a fraction is to place a minus sign above the negative characteristic; for example, $\overline{1}.7059$. Show the two other ways to write this expression:

____

____

- - - - - - - - - - - - - - - - - -

-1 + .7059

.7059 - 1

21. You may recall that fractions or divisions of common numbers change to subtractions if converted into logs. Conversely, if logs are converted to common numbers, subtractions change to fractions or divisions; for example,

antilog .7059 - 1 = antilog .7059 — antilog 1

or $\dfrac{\text{antilog } .7059}{\text{antilog } 1} = \dfrac{\_\_\_\_}{10}$.

- - - - - - - - - - - - - - - - - -

$\dfrac{5.08}{10}$

22. antilog $\overline{1}.7059 = \dfrac{5.08}{10} =$ ________

- - - - - - - - - - - - - - - - - -

.508

23. PROBLEM: Find the antilog $\overline{3}.7782$:

____

$\frac{\text{antilog} \_\_\_\_\_\_}{\text{antilog}}$

- - - - - - - - - - - - - - - - - -

$\frac{.7782}{3}$

24. antilog $\bar{3}.7782 = \frac{\text{antilog } .7782}{\text{antilog } 3} =$ __________

- - - - - - - - - - - - - - - - - -

$\frac{6}{1000}$

25. antilog $\bar{3}.7782 = \frac{\text{antilog } .7782}{\text{antilog } 3} = \frac{6}{1000} =$ __________

- - - - - - - - - - - - - - - - - -

.006

26. PROBLEM: Find the antilog $\bar{4}.6561$:

$\frac{\text{antilog } .6561}{\text{antilog}} =$ __________ = __________

- - - - - - - - - - - - - - - - - -

$\frac{.6561}{4} = \frac{4.53}{10{,}000} = .000453$

27. PROBLEM. Find the antilog $\bar{3}.9795$:

$\frac{\text{antilog} \_\_\_\_\_\_}{\text{antilog}} =$ __________ = __________

- - - - - - - - - - - - - - - - - -

$$\frac{.9795}{3} = \frac{9.54}{1000} = .00954$$

28. antilog $\bar{6}.9571$ = __________

- - - - - - - - - - - - - - - - - - -

.00000906

Antilog $\bar{6}.9571$ equals .00000906. This is an unwieldy number. Generally it would be written 9.06×10^{-6}, which is exactly the same number but in a different form. We express unwieldy numbers like .00000906 in powers of 10 as shown above.

29. Write the antilog $\overline{18}.9571$ in powers of 10. __________

- - - - - - - - - - - - - - - - - - -

9.06×10^{-18}

30. We can also express positive numbers that are unwieldy by using powers of 10; for example, antilog 10.9571 = 9.06×10^{10} rather than 90,600,000,000.

Write antilog 18.9571 in powers of 10. __________

- - - - - - - - - - - - - - - - - - -

9.06×10^{18}

31 It is very important to be extremely careful in observing the signs used in mathematics. Note that when we write

antilog $\overline{18}.9571 = 9.06 \times 10^{-18}$

antilog $18.9571 = 9.06 \times 10^{18}$

the only difference is in the power designation, but what a difference there would be if we were to write out the results! Compare the difference:

antilog $\overline{18}.9571$ = .000000000000000000906

antilog 18.9571 = 9060000000000000000.

32. Common logarithms are a form of powers of ______ (number) in which the ______________ (base, exponent) is assumed and left out.

- - - - - - - - - - - - - - - - - - -

10

base

33. When using logs, multiplications of numbers become simple _______________.

- - - - - - - - - - - - - - - - - - -

additions

34. When using logs, divisions of numbers become simple _______________.

- - - - - - - - - - - - - - - - - - -

subtractions

35. To find the antilog 2.4771 we ______________ the antilog 2 by the antilog .4771.

- - - - - - - - - - - - - - - - - - -

multiply

36. To find the antilog $\bar{3}.7782$ we ____________ the antilog ____________ by the antilog __________.

- - - - - - - - - - - - - - - - - - -

divide

.7782

3

SELF-TEST

1. antilog 3.1679 = ____________

2. antilog 4.2651 = ____________

3. antilog 2.4411 = ____________

4. antilog 1.6297 = ____________

5. antilog 3.5761 = ____________

6. antilog $\bar{2}.6873$ = ____________

7. antilog $\bar{1}.8033$ = ____________

8. antilog $\bar{3}.8665$ = ____________

9. antilog $\bar{4}.9006$ = ____________

10. antilog $\bar{2}.9471$ = ____________

Show antilogs of characteristics in exponential form.

11. antilog 18.7361 = ____________

12. antilog 15.6780 = ____________

13. antilog 9.9327 = ______________

14. antilog 12.0341 = ______________

15. antilog 6.5278 = ______________

16. antilog $\overline{11}$.4541 = ______________

17. antilog $\overline{15}$.5285 = ______________

18. antilog $\overline{9}$.6741 = ______________

19. antilog $\overline{6}$.7943 = ______________

20. antilog $\overline{5}$.9152 = ______________

Answers

1. 1,472

2. 18,410

3. 276.1

4. 42.63

5. 3,768

6. .04868

7. .6357 or .6358

8. .007353 or .007354

9. .0007953 or .0007954

10. .08854 or .08855

11. 5.446×10^{18}

12. 4.764×10^{15}

13. 8.563×10^{9} or 8.564×10^{9}

14. 1.082×10^{12}

15. 3.371×10^{6} or 3.372×10^{6}

16. 2.845×10^{-11}

17. 3.377×10^{-15}

18. 4.722×10^{-9}

19. 6.227×10^{-6}

20. 8.225×10^{-5} or 8.226×10^{-5}

UNIT TEN

Multiplication by Use of Logarithms

1. We have already learned that by using logs multiplication becomes simple ________________.

- - - - - - - - - - - - - - - - - - -

addition

2. If, for instance, we wanted to multiply 193 x 654, we would have two alternate methods. First we could use the long method (arithmetical):

```
     193
x    654
--------
     772
    965
  1158
--------
  126222
```

Or we could use our knowledge of logarithms and change the operation from a multiplication to an addition by the following steps:

Step 1 Find the logs of the numbers to be multiplied:

log 193 = 2.2856

log 654 = 2.8156

Step 2 Add these logs.

    2.2856
  + 2.8156
    5.1012

Step 3 Find the antilog of the sum.

antilog 5.2012 = 1.262×10^5 = 126,200

Thus by arithmetic we found that the product was 126,222 and by the use of logs the product was 126,200. You will note that there is a slight difference of 22. The degree of accuracy necessary depends on the purpose of the calculations; however, when dealing with large numbers, this minor difference can be ignored for most purposes.

3. PROBLEM: By the use of logs multiply 3765 x 8567.

log 3765 = ____________

log 8567 = ____________

- - - - - - - - - - - - - - - - - - -

3.5758

3.9329

4. 3.5758
  +3.9329

- - - - - - - - - - - - - - - - - - -

7.5087

5. antilog 7.5087 = ____________ x 10——

- - - - - - - - - - - - - - - - - -

3.226 x 10^7

6. antilog 7.5087 = ____________

- - - - - - - - - - - - - - - - - -

32, 260, 000

7. PROBLEM: By the use of logs find the product of 867 x 123:

log 867 = ____________

log 123 = ____________

antilog = ____________ = ______________

- - - - - - - - - - - - - - - - - -

2.9380

2.0899

5.0279 = 106, 600

8. PROBLEM. By the use of logs find the product of 177 x 523:

log 177 = ____________

log 523 = ____________

antilog = ____________ = ______________

- - - - - - - - - - - - - - - - - -

2.2480

2.7185

4.9665 = 92, 580

9. We can also multiply whole numbers and fractions. Remember, however, that mantissas are always positive, whereas characteristics can be either positive or negative. It ___________ (is, is not) possible to have a negative mantissa and a positive characteristic.

- - - - - - - - - - - - - - - - - - -

is not

10. It ___________ (is, is not) possible to have a positive mantissa and a negative characteristic.

- - - - - - - - - - - - - - - - - - -

is

11. PROBLEM. By the use of logs find the product of 177 x .523:

Our first step is the same as we used in multiplying whole numbers. We find the logs of the two numbers:

log 177 = ___________

log .523 = ___________

- - - - - - - - - - - - - - - - - - -

2.2480

$\bar{1}.7185$

12. You will recall in Set 7 we learned that any number below 1 is a fraction and therefore a negative log. Further, you will recall that negative logs are expressed by placing a minus sign above the characteristic. Therefore the log .523 is expressed as $\bar{1}.7185$.

The second step in multiplying whole numbers and fractions is similar to the process we learned in multiplying whole numbers. Logs that have negative characteristics and positive mantissas are added algebraically. (Adding algebraically means adding plus and minus quantities.)

$$\begin{aligned} \log 177 &= 2.2480 \\ \log .523 &= \bar{1}.7185 \\ \hline & 1.9665 \end{aligned}$$

Remember that the minus sign above the characteristic makes the characteristic negative but not the mantissa.

13. In the above operation we added two logs algebraically and obtained 1.9665. Now we can proceed as we did in multiplying whole numbers. We find the antilog:

antilog 1.9665 = 9.258 x ______ = __________

- - - - - - - - - - - - - - - - - - -

10^1 = 92.58

14. PROBLEM. By the use of logs find the product of 639 x .037.

Find the logs of the two numbers

log 639 = __________

log .037 = __________

- - - - - - - - - - - - - - - - - - -

2.8055

$\bar{2}.5682$

15. Add the two logs algebraically. First add the mantissas:

$$\begin{array}{r} .8055 \\ \underline{.5682} \\ 1.3737 \end{array}$$

You will notice that in this addition we ended up with an extra digit, the 1 that precedes the decimal. This 1 is important because we must carry it over to our characteristics. The rule is that any additional digit resulting from adding mantissas is carried over to the characteristics and added algebraically to the characteristics. If we carry the extra digit to the characteristics and add algebraically, we will get the following:

$$\begin{array}{r} 2.8055 \\ \underline{\bar{2}.5682} \\ 1.3737 \end{array}$$

The next step is to find the antilog:

antilog 1.3737 = __________ x ______ = __________

$2.364 \times 10^1 = 23.64$

16. Add the following numbers algebraically:

(a) $\begin{array}{r} 2.5678 \\ \underline{\bar{2}.5321} \end{array}$ (b) $\begin{array}{r} 2.1461 \\ \underline{\bar{1}.7404} \end{array}$ (c) $\begin{array}{r} 5.2345 \\ \underline{\bar{2}.1234} \end{array}$ (d) $\begin{array}{r} 2.1987 \\ \underline{\bar{3}.2456} \end{array}$

(e) $\begin{array}{r}2.9654\\ \underline{\bar{2}.2222}\end{array}$ (f) $\begin{array}{r}1.1357\\ \underline{\bar{2}.9654}\end{array}$ (g) $\begin{array}{r}1.2121\\ \underline{1.7879}\end{array}$ (h) $\begin{array}{r}2.5678\\ \underline{\bar{4}.5321}\end{array}$

- - - - - - - - - - - - - - - - - -

(a) 1.0999 (b) 1.8865 (c) 3.3579 (d) $\bar{1}.4443$

(e) 1.1876 (f) 0.1011 (g) 1.0000 (h) $\bar{1}.0999$

17. By the use of logs find the product of 345 x .000578 = ________.

- - - - - - - - - - - - - - - - - -

log 345 = 2.5378

log .000578 = $\bar{4}.7619$

antilog = $\bar{1}.2997$ = 1.994 x 10^{-1} = .1994

18. By the use of logs find the product of 753.5 x .0006788 = ________.

- - - - - - - - - - - - - - - - - -

log 753.5 = 2.8771

log .0006788 = $\bar{4}.8318$

antilog = $\bar{1}.7089$ = 5.116×10^{-1} = .5116

19. For a series of multiplications logs are especially useful provided that absolute accuracy is not needed; for instance, multiply

56.7 x 78.5 x 7.5 x 7.656 x 8534 x 6794 = ?

We solve this type of problem exactly as we did the others.

log 56.7 = __________

log 78.5 = __________

log 7.5 = __________

log 7.656 = __________

log 8534 = __________

log 6794 = __________

antilog __________ = ________ x 10—

- - - - - - - - - - - - - - - - - -

1.7536

1.8949

.8751

.8840

3.9311

3.8322

13.1709 = 1.482×10^{13}

If the answer contains a large exponent (regardless of the sign), it is best to leave it in that form. Would you attempt to write out 7.8 x 10^{36}? Of course not. Not only would it be a tedious job to write it

7,800,000,000,000,000,000,000,000,000,000,000,000,000

but it is practically impossible to read!

20. By the use of logs multiply

77.5 x 365 x 977.5 x .00356 x 7.835 x .9395 = ______

- - - - - - - - - - - - - - - - - -

log 77.5	=	1.8893
log 365	=	2.5623
log 977.5	=	2.9901
log .00365	=	$\bar{3}.5514$
log 7.835	=	.8941
log .9395	=	$\bar{1}.9729$
antilog	=	5.8601 = 7.246 x 10^{5}

21. By the use of logs multiply

8977 x 5698 x 4785 x 8954 x 7532 = ________

- - - - - - - - - - - - - - - - - -

log 8977	=	3.9531		
log 5698	=	3.7552		
log 4785	=	3.6799		
log 8954	=	3.9520		
log 7532	=	3.8769		
antilog	=	19.2171	=	1.649×10^{19}

22. By the use of logs multiply

.00376 x .0005983 x .007985 x .004954 x .07643
= ________

- - - - - - - - - - - - - - - - - -

log .00376	=	$\bar{3}.5752$
log .0005983	=	$\bar{4}.7769$
log .007985	=	$\bar{3}.9023$
log .004954	=	$\bar{3}.6950$

log .07643 $= \bar{2}.8831$

antilog $= \overline{12}.8325 = 6.8 \times 10^{-12}$

SELF-TEST

By the use of logs perform the following operations. Leave the antilogs of the characteristics in exponential form whenever the number becomes unwieldy.

1. 1.598 x 6.715 = ____________
2. 3.775 x 4.936 = ____________
3. 85.72 x 13.25 = ____________
4. 2.541 x 9853 = ____________
5. 4.679 x 7,596 = ____________
6. 8,779 x 75.88 = ____________
7. 3,521 x 656.8 = ____________
8. 5,839 x 7,496 = ____________
9. 6785 x 5315 x 7589 x 1755 x 980.5 = __________
10. 17.56 x 637.9 x 186.5 x 899.7 x 15.87 = __________
11. 9891 x 156.7 x 1785 x 831.7 x 776.5 = __________
12. .8739 x .1673 x .5493 x .9785 = __________
13. 7395 x 855.6 x .8945 x 7445 x .1735 = __________
14. .003754 x .07856 x .0009416 x .5391 = __________

15. $53.99 \times 5.326 \times 10^{-5} \times 1.819 \times 10^{7} \times 5.763 \times 10^{-12}$

= ____________

Answers

1.	log 1.598	=	.2036	
	log 6.715	=	.8270	
	antilog		1.0306	= 10.73
2.	log 3.775	=	.5769	
	log 4.936	=	.6933	
	antilog		1.2702	= 18.63
3.	log 85.72	=	1.9331	
	log 13.25	=	1.1222	
	antilog		3.0553	= 1,136
4.	log 2.541	=	.4050	
	log 9853	=	3.9935	
	antilog		4.3985	= 25,030 or 25,040
5.	log 4.679	=	.6701	
	log 7596	=	3.8805	
	antilog	=	4.5506	= 35,530
6.	log 8779	=	3.9434	
	log 75.88	=	1.8802	
	antilog		5.8236	= 6.661×10^{5}

7.	log 3521	=	3.5466	
	log 656.8	=	2.8174	
	antilog		6.3640	$= 2.312 \times 10^{6}$
8.	log 5839	=	3.7664	
	log 7496	=	3.8749	
	antilog		7.6413	$= 4.378 \times 10^{7}$
9.	log 6785	=	3.8315	
	log 5315	=	3.7255	
	log 7589	=	3.8802	
	log 1755	=	3.2442	
	log 980.5	=	2.9914	
	antilog		17.6728	$= 4.708 \times 10^{17}$
10.	log 17.56	=	1.2445	
	log 637.9	=	2.8047	
	log 186.5	=	2.2707	
	log 899.7	=	2.9541	
	log 15.87	=	1.2007	
	antilog	=	10.4747	$= 2.983 \times 10^{10}$ or 2.984×10^{10}
11.	log 9891	=	3.9952	
	log 156.7	=	2.1951	
	log 1785	=	3.2516	
	log 831.7	=	2.9200	
	log 776.5	=	2.8903	
	antilog		15.2522	$= 1.737 \times 10^{15}$

12.	log .8739	=	$\bar{1}.9414$		
	log .1673	=	$\bar{1}.2235$		
	log .5493	=	$\bar{1}.7398$		
	log .9785	=	$\bar{1}.9907$		
	antilog		$\bar{2}.8954$	=	.0786
13.	log 7395	=	3.8689		
	log 855.6	=	2.9323		
	log .8945	=	$\bar{1}.9515$		
	log 7445	=	3.8719		
	log .1735	=	$\bar{1}.2392$		
	antilog		9.8638	=	7.308×10^{9} or 7.309×10^{9}
14.	log .003754	=	$\bar{3}.5745$		
	log .07856	=	$\bar{2}.8952$		
	log .0009416	=	$\bar{4}.9739$		
	log .5391	=	$\bar{1}.7317$		
	antilog		$\bar{7}.1753$	=	1.497×10^{-7}
15.	log 53.99	=	1.7323		
	log 5.326	=	.7264		
	log 10^{-5}	=	$\bar{5}.0000$		
	log 1.819	=	.2598		
	log 10^{7}	=	7.0000		
	log 5.763	=	.7606		
	log 10^{-12}	=	$\overline{12}.0000$		
	antilog		$\bar{7}.4791$	=	3.013×10^{-7} or 3.014×10^{-7}

UNIT ELEVEN

Division by Use of Logarithms

1. Division of common numbers is reduced to simple ________________ if logs are used.

- - - - - - - - - - - - - - - - - - -

subtraction

2. As in multiplying by using logs, the first step in dividing by using logs is to find the logs of the numbers. But then we subtract rather than add; for instance, divide 4365 by 365.

log 4365	=	3.6400
log 365	=	-2.5623

- - - - - - - - - - - - - - - - - - -

1.0777

Now all we have to do is find the antilog 1.0777 which is ______________. This number is the result of the division.

- - - - - - - - - - - - - - - - - - -

11.96

3. By the use of logs find the result of division 855 ÷ 365.

log 855 = ________

log 365 = ________

antilog ________ = ________

- - - - - - - - - - - - - - - - - -

2.9320

2.5623

.3697 = 2.343 or 2.342

4. By the use of logs find the result of division 935 ÷ 87 = ________

- - - - - - - - - - - - - - - - - -

log 935 = 2.9708

log 87 = -1.9395

antilog 1.0313 = 1.075 (or 1.074) x 10^1

= 10.75 or 10.74

5. Up to this point we have dealt with divisions in which we divided a larger number by a smaller number. Suppose, however, we were asked to solve this problem:

87 ÷ 935

The result of division 87 ÷ 935 will be a ______________ (whole number, fraction).

- - - - - - - - - - - - - - - - - -

fraction

6. As we have already learned, fractions will have a negative __________________ (characteristic, mantissa).

- - - - - - - - - - - - - - - - - -

characteristic

7. As in addition, characteristics and mantissas are treated as separate entities when performing subtraction of logs.

2.3979	3.9047	2.9047
-1.3222	-2.4771	-2.9031

- - - - - - - - - - - - - - - - - -

1.0757

1.4276

0.0016

8. Let's look at this problem:

$$\begin{array}{r} 3.3979 \\ -1.9222 \\ \hline ? \end{array}$$

As in the case of adding mantissas and characteristics, we must carry over the extra digit into the characteristics. This time, however, we <u>subtract</u> the extra digit from our upper characteristic.

$$\begin{array}{r} 2 \\ \not{3}.3979 \\ -1.9222 \\ \hline 1.4757 \end{array}$$

9. Find the answers:

5.6789	8.2134	2.1987	2.0000
-1.9876	-2.8213	-1.5432	-1.0009

- - - - - - - - - - - - - - - - - -

4	7	1	1
$\not{5}$.6789	$\not{8}$.2134	$\not{2}$.1987	$\not{2}$.0000
-1.9876	-2.8213	-1.5432	-1.0009
3.6913	5.3921	0.6555	0.9991

10. In algebra, when we subtract a larger number from a smaller number, the result is a ________________ number.

- - - - - - - - - - - - - - - - - -

negative

11. Therefore, when subtracting characteristics and mantissas, we observe the same algebraic rule; for instance,

$$\begin{array}{r} 2.3979 \\ -3.3222 \\ \hline \bar{1}.0757 \end{array} \qquad \begin{array}{r} 3.9047 \\ -4.4771 \\ \hline \bar{1}.4276 \end{array} \qquad \begin{array}{r} 2.9047 \\ -3.9031 \\ \hline \end{array}$$

- - - - - - - - - - - - - - - - - -

$\bar{1}.0016$

12. When we have a "carry" number, we treat it in exactly the same manner:

$$\begin{array}{r} 4 \\ \not{5}.6789 \\ -6.9876 \\ \hline \bar{2}.6913 \end{array}$$

Find the answers:

$$\begin{array}{r} 5.6789 \\ -7.9876 \\ \hline \end{array} \qquad \begin{array}{r} 8.2134 \\ -9.8213 \\ \hline \end{array} \qquad \begin{array}{r} 2.1987 \\ -5.5432 \\ \hline \end{array} \qquad \begin{array}{r} 2.0000 \\ -3.0009 \\ \hline \end{array}$$

- - - - - - - - - - - - - - - - - -

$$\begin{array}{r} 4 \\ \not{5}.6789 \\ -7.9876 \\ \hline \bar{3}.6913 \end{array} \qquad \begin{array}{r} 7 \\ \not{8}.2134 \\ -9,8213 \\ \hline \bar{2}.3921 \end{array} \qquad \begin{array}{r} 1 \\ \not{2}.1987 \\ -5.5432 \\ \hline \bar{4}.6555 \end{array} \qquad \begin{array}{r} 1 \\ \not{2}.0000 \\ -3.0009 \\ \hline \bar{2}.9991 \end{array}$$

13. We are now ready to solve our problem: 87 ÷ 935

log 87 = 1.9395

log 935 = -2.9708

antilog $\bar{2}.9687$ = 9.035 x 10^{-2} = .09305

14. By the use of logs find the result of division 765 ÷ 899.

log 765 = ________

log 899 = - ________

antilog ________ = ________ x ________

= ________

- - - - - - - - - - - - - - - - - -

3.8837

3.9538

$\bar{1}.9299$ = 8.51 x 10^{-1}

= .851

15. By the use of logs find the result of division .0004567 ÷ 5450.

log .0004567 = $\bar{4}.6597$

log 5450 = -3.7364

antilog $\bar{8}.9233$ = 8.38 x 10^{-8}

= .0000000838

As already stressed, we treat characteristics and mantissas separately and perform algebraic operations exactly as practiced. Special care must be exercised to keep track of signs in arriving at results.

16. Find the answers:

$$\begin{array}{cccc} \bar{5}.6789 & 8.2134 & \bar{2}.1987 & 2.0000 \\ \underline{-1.9876} & \underline{-\bar{2}.8213} & \underline{-\bar{1}.5432} & \underline{-1.0009} \end{array}$$

- - - - - - - - - - - - - - - - - -

$$\begin{array}{cccc} \bar{7}.6913 & 9.3921 & \bar{2}.6555 & 2.9991 \end{array}$$

17. By the use of logs find the result of division .0356 ÷ .00375. ____________

- - - - - - - - - - - - - - - - - -

$$\begin{array}{lcl} \text{log } .0356 & = & \bar{2}.5514 \\ \text{log } .00375 & = & \underline{-\bar{3}.5740} \\ \text{antilog} & & 0.9774 = 9.492 \end{array}$$

18. By the use of logs find the result of division .00536 ÷ .0589. ____________

- - - - - - - - - - - - - - - - - -

log .00536	=	$\bar{3}.7292$		
log .0589	=	$-\bar{2}.7701$		
antilog		$\bar{2}.9591$	$= 9.102 \times 10^{-2}$	= .09102

19. By the use of logs find the result of division .000175 ÷ .396.

____

- - - - - - - - - - - - - - - - - - -

log .000175	=	$\bar{4}.2430$		
log .3960	=	$-\bar{1}.5977$		
antilog		$\bar{4}.6453$	$= 4.419 \times 10^{-4}$	= .0004419

SELF-TEST

1. 7369 ÷ 654 = _______________
2. 8563 ÷ 371 = _______________
3. 1956 ÷ 339 = _______________
4. 5963 ÷ 763 = _______________
5. 7869 ÷ 211 = _______________
6. 4939 ÷ 57.88 = _______________
7. 6738 ÷ 95.43 = _______________
8. 3584 ÷ 7853 = _______________

9. .7984 ÷ 7873 = ____________

10. .003945 ÷ .05676 = ____________

11. $\frac{5873}{7875}$ = ____________

12. $\frac{.1285}{.3546}$ = ____________

13. $\frac{.007865}{.09876}$ = ____________

14. $\frac{.7987}{.0008433}$ = ____________

15. $\frac{.003785}{.0004839}$ = ____________

Answers

1. log 7369 = 3.8674
log 654 = -2.8156
antilog 1.0518 = 11.27

2. log 8563 = 3.9327
log 371 = -2.5694
antilog 1.3633 = 23.08 or 23.09

3.	log 1956	=	3.2913	
	log 339	=	-2.5302	
	antilog		.7611	= 5.769
4.	log 5963	=	3.7754	
	log 763	=	-2.8825	
	antilog		.8929	= 7.813 or 7.814
5.	log 7869	=	3.8959	
	log 211	=	-2.3243	
	antilog		1.5716	= 37.29
6.	log 4939	=	3.6936	
	log 57.88	=	-1.7625	
	antilog		1.9311	= 85.33 or 85.34
7.	log 6738	=	3.8285	
	log 95.43	=	-1.9796	
	antilog		1.8489	= 70.61 or 70.62
8.	log 3584	=	3.5544	
	log 7853	=	-3.8951	
	antilog		$\bar{1}.6593$	= .4563
9.	log .7984	=	$\bar{1}.9022$	
	log 7873	=	-3.8962	
	antilog		$\bar{4}.0060$	= 1.014×10^{-4} = .0001014
10.	log .003945	=	$\bar{3}.5960$	
	log .05676	=	$-\bar{2}.7541$	
	antilog		$\bar{2}.8419$	= .06948

11. log 5873 = 3.7688
log 7875 = -3.8963
antilog $\bar{1}.8725$ = .7458 or .7459

12. log .1285 = $\bar{1}.1089$
log .3546 = $-\bar{1}.5497$
antilog $\bar{1}.5592$ = .3624

13. log .007865 = $\bar{3}.8957$
log .09876 = $-\bar{2}.9946$
antilog $\bar{2}.9011$ = .07963 or .07964

14. log .7987 = $\bar{1}.9024$
log .0008433 = $-\bar{4}.9260$
antilog 2.9764 = 947.2 or 947.3

15. log .003785 = $\bar{3}.5581$
log .0004839 = $-\bar{4}.6847$
antilog .8734 = 7.471 or 7.472

UNIT TWELVE

Raising to a Power by the Use of Logarithms

1. In operations in which logs are used multiplication is reduced to _______________.

- - - - - - - - - - - - - - - - - -

addition

2. For example,

$(100)^3 = 100 \times 100 \times 100 = 1,000,000$

We can also do this by using logs:

$\log (100)^3 = \log 100 + \log 100 + \log 100$

If $\log 100 = 2$, then

$\log (100)^3 = 2 + 2 + 2 =$ ________

- - - - - - - - - - - - - - - - - -

6

3. antilog 6 = ___________

- - - - - - - - - - - - - - - - - -

1,000,000

4. We can also write the operation in a different way.

$\log (100)^3 = \log 100 + \log 100 + \log 100$

$= (\log 100)3$

$= (2)\ 3 =$ _______

- - - - - - - - - - - - - - - - - - -

6

Thus we get the same answer either way.

$(100)^3 = 100 \times 100 \times 100 = 1,000,000$

$\log (100)^3 = (\log 100)3 = (2)3 = 2 \times 3 = 6$

antilog 6 = 1,000,000

5. $(100)^3 = 1,000,000$ is a fairly easy operation and we don't need logarithms to do it. But what about an operation like $(587)^5$? If we can say that $\log (100)^3 = (\log 100)3$, we can say the same thing about any other number raised to any power; for instance,

$\log (587)^5 = (\log 587)5$

If log 587 = 2.7686, then

$(\log 587)5 = 2.7686 \times 5 =$ _________

- - - - - - - - - - - - - - - - - - -

13.8430

6. Now all we have to do is find the antilog 13.8430 to determine our number for $(587)^5$.

antilog 13.8430 (expressed in powers of 10) = _________

x 10___

- - - - - - - - - - - - - - - - - - -

6.967×10^{13}

$\log (587)^5 = (\log 587)5 = 2.7686 \times 5 = 13.8430$

antilog 13.8430 = 6.996 or 6.967 x 10

By using logs we solved our problem in less than two lines. Now compare this with the long method.

587 x 587 x 587 x 587 x 587

Do you feel like trying it the long way. If so, good luck, and we'll see you next week! If you don't want to waste time, use logs.

7. When using logs, multiplication of numbers is reduced to __________ and raising numbers to powers is reduced to ____________.

- - - - - - - - - - - - - - - - - - -

addition

multiplication

8. Most of the time you will see $\log(587)^5 = (\log 587)5$ written as 5 log 587; you would read this as "five times the log of 587."

$\log(786)^2 = 2 \log 786$

$\log(365)^4 =$ ____ log 365

$(598)^{17} =$ __________

- - - - - - - - - - - - - - - - - - -

4

17 log 598

9. $\log(681)^7 = 7 \log 681 = 7(\underline{\hspace{6em}}) = \underline{\hspace{5em}}$

- - - - - - - - - - - - - - - - - - -

(2.8331) 19.8317

10. antilog 19.8317 = ________ x _____

- - - - - - - - - - - - - - - - - - -

6.787 (or 6.788) $\times 10^{19}$

11. $\log(371)^{20}$ = _____ log _____ = _____ (________)

= ____________

- - - - - - - - - - - - - - - - - - -

$20 \log 371 = 20(2.5694) = 51.3880$

12. antilog 51.3880 = ________ x _____

- - - - - - - - - - - - - - - - - - -

2.443×10^{51}

13. By the use of logs perform the following operation:

$(5.864)^7$ = __________

- - - - - - - - - - - - - - - - - - -

$7 \log 5.864 = 7(.7682) = 5.3774$

antilog $5.3774 = 2.384 \times 10^5$ or 2.385×10^5

14. By the use of logs perform the following operation (be careful!):

$(.5864)^7 = $ ____________

- - - - - - - - - - - - - - - - - -

$7 \log .5864 = 7(\bar{1}.7682) = \bar{2}.3774$

$\text{antilog } \bar{2}.3774 = .02384 \text{ or } .02385$

Again the characteristic and the mantissa are treated as separate entities. The last problem could be written

$$(.5864)^7 = 7 \log .5864 = 7(-1 + .7682) = -7 + 5.3774$$
$$= -2 + 3774$$
$$= \bar{2}.3774, \text{ etc.}$$

15. By the use of logs perform the following operations:

$(.003971)^9 = $ ____________

$(.0005693)^{11} = $ ____________

- - - - - - - - - - - - - - - - - -

$$9 \log .003971 = 9(\bar{3}.5989)$$
$$= \overline{22}.3901$$
$$= 2.455 \times 10^{-22}$$

$$11 \log .0005693 = 11(\bar{4}.7553)$$
$$= \overline{36}.3083$$
$$= 2.034 \times 10^{-36}$$

16. By the use of logs, multiplication is reduced to ____________; division is reduced to ________________; and raising to a power is reduced to __________________.

- - - - - - - - - - - - - - - - - -

addition

subtraction

multiplication

SELF-TEST

1. $(5877)^5$ = ____________

2. $(3563)^9$ = ____________

3. $(8735)^{12}$ = ____________

4. $(736.9)^{15}$ = ____________

5. $(6.456)^{18}$ = ____________

6. $(2)^{63}$ = ___________

7. $(1.037)^{13}$ = ___________

8. $(.3794)^{12}$ = ___________

9. $(.00984)^{5}$ = ___________

10. $(.0005396)^{7}$ = ___________

Answers

1. 5 log 5877 = 5(3.7691) = 18.8455
   antilog 18.8455 = 7.006×10^{18}

2. 9 log 3563 = 9(3.5518) = 31.9662
   antilog 31.9662 = 9.252×10^{31} or 9.253×10^{31}

3. 12 log 8735 = 12(3.9412) = 47.2944
   antilog 47.2944 = 1.969×10^{47}

4. 15 log 736.9 = 15(2.8674) = 43.0110
   antilog 43.0110 = 1.026×10^{43}

5. 18 log 6.456 = 18(.8100) = 14.5800
   antilog 14.5800 = 3.802×10^{14}

6. 63 log 2 = 63(3010) = 18.9630
   antilog 18.9630 = 9.184×10^{18} or 9.185×10^{18}

7. 13 log 1.037 = 13(.0157) = .2041
   antilog .2041 = 1.600 or just 1.6

8. 12 log .3794 = $12(\bar{1}.5791)$ = $\bar{6}.9492$
   antilog $\bar{6}.9492$ = 8.896×10^{-6} or 8.897×10^{-6}

9. $5 \log .00984 = 5(\bar{3}.9930) = \overline{11}.9650$

antilog $\overline{11}.9650 = 9.226 \times 10^{-11}$ or 9.227×10^{-11}

10. $7 \log .0005396 = 7(\bar{4}.7321) = \overline{23}.1247$

antilog $\overline{23}.1247 = 1.332 \times 10^{-23}$ or 1.333×10^{-23}

UNIT THIRTEEN

Extracting Roots by the Use of Logarithms

1. We know that $10^2 = 10 \times 10 = 100$

$(10)^4 =$ ______________ = __________

- - - - - - - - - - - - - - - - - - -

$10 \times 10 \times 10 \times 10 = 10,000$

2. We also remember from our work in algebra that we can find the root of a number; for instance,

$$\sqrt[2]{144} = \begin{array}{r|rr} & 1 & 2 \\ \sqrt[2]{} & 1 & 44 \\ & 1 & \\ 22 & & 44 \\ 2 & & 44 \\ \hline \end{array}$$

$$\sqrt[2]{60025} = \begin{array}{r|rrr} & 2 & 4 & 5 \\ \sqrt[2]{} & 6 & 00 & 25 \\ & 4 & & \\ 44 & 2 & 00 & \\ 4 & 1 & 76 & \\ \hline 485 & & 24 & 25 \\ 5 & & 24 & 25 \\ \hline \end{array}$$

Proof:

$$\begin{array}{r} 12 \\ \times 12 \\ \hline 24 \\ 12 \\ \hline 144 \end{array}$$

Proof:

$$\begin{array}{r} 245 \\ \times 245 \\ \hline 1225 \\ 980 \\ 490 \\ \hline 60025 \end{array}$$

We can also find roots by using logs.

3. If we were asked to solve this problem.

$\sqrt[2]{10,000}$

we could do it by using logs in this way:

$\log \sqrt[2]{10,000} = (\log 10,000) \div 2; \ \log 10,000 = 4$

$= 4 \div 2 = 2$

$=$ antilog $2 = 100$, therefore:

100 is the square root of 10,000.

Proof: 100 x 100 = 10,000.

4. Using logs, find the cube root of 1,000,000.

$\log \sqrt[3]{1,000,000}$

$(\log 1,000,000) \div 3 =$ _____ $\div 3 =$ _____

antilog _____ = _____

- - - - - - - - - - - - - - - - - - - -

$6 \div 3 = 2$

$2 = 100$

5. Using logs, find the fifth root of 100,000.

$\log \sqrt[5]{100,000}$

$(\log 100,000) \div$ _____ = _____ $\div$ _____ = _____

antilog _____ = _____

- - - - - - - - - - - - - - - - - - - -

5 = 5 ÷ 5 = 1

1 = 10

6. Using logs, find the square root of 2856.

log $\sqrt[2]{2856}$

(log 2856) ÷ 2 = ________ ÷ 2 = ___________

antilog ____________ = ____________

- - - - - - - - - - - - - - - - - - -

3.4557 ÷ 2 = 1.7278

1.7278 = 53.44

7. Using logs, find the fifth root of 2856.

log $\sqrt[5]{2856}$

(log 2856) ÷ _____ = 3.4557 ÷ _____ = ___________

antilog ____________ = _____________

- - - - - - - - - - - - - - - - - - -

5 = 3.4557 ÷ 5 = .6911

.6911 = 4.910

8. Using logs, find the seventh root of 7856.

log $\sqrt[7]{7856}$ = __________

antilog ____________ = ___________

- - - - - - - - - - - - - - - - - - -

(log 7856) ÷ 7 = 3.8952 ÷ 7 = .5565

.5565 = 3.602

9. Using logs, find the ninth root of 7856.

log $\sqrt[9]{7856}$ = ____________

- - - - - - - - - - - - - - - - - - -

(log 7856) ÷ 9 = 3.8952 ÷ 9 = .4328

antilog .4328 = 2.709

10. By the use of logs multiplication is reduced to ____________.

- - - - - - - - - - - - - - - - - - -

addition

11. By the use of logs division is reduced to ____________.

- - - - - - - - - - - - - - - - - - -

subtraction

12. By the use of logs raising to a power is reduced to ____________.

- - - - - - - - - - - - - - - - - - -

multiplication

13. By the use of logs extracting a root is reduced to __________.

- - - - - - - - - - - - - - - - - -

division

14. Now that we know how to find the root of a whole number by using logs, let's turn our attention to finding roots of fractions by using logs. Let's assume that we were asked to find the root of

$\log \sqrt[2]{.81}$

We would start exactly as we have been:

$\log \sqrt[2]{.81} = (\log .81) \div 2 = \bar{1}.9085 \div 2$

Before going on, let's take a second look at the log number.

Log .81 = $\bar{1}.9085$. This number has a negative characteristic and a positive mantissa. If we were to calculate $\log \sqrt{.81}$ which is $\bar{1}.9085 \div 2$, as we have been doing to find the root of whole numbers, we would be wrong. Let's check it. By the standard method $\sqrt{.81} = .9$. Using logs and proceeding without regard to signs, we calculate as follows:

$\log \sqrt[2]{.81} = \bar{1}.9085 \div 2 = .95425$ or $.9543$

antilog .9543 = 9.0

which is obviously wrong.

15. We must find a way to make the division correct. A short detour might be helpful. Take a close look at the figure below:

$-1 = -2 + 1$

Both sides of this equation are _________________ (the same, not the same).

- - - - - - - - - - - - - - - - - -

the same

16. $-1 \div 2 = -.5$ and $(-2 + 1) \div 2 = -.5$

Both equations give the same result; therefore

$-1 \div 2 - (-$ _____ $+$ _____$) \div 2$

- - - - - - - - - - - - - - - - - -

$(-2 + 1)$

17. We return to our problem, which was

$\log {}^{2}.81 = (\log .81) \div 2 = \bar{1}.9085 \div 2$

As we know, we cannot divide a number with a negative characteristic and a positive mantissa, but we can change the characteristic so that it can be divided by 2:

$\bar{1}.9085 \div 2$ can be changed to $\bar{2} + 1.9085 \div 2$

We have added a -1 to the characteristic and a +1 to the mantissa.

Now we have a chracteristic of two parts, $\bar{2} + 1$, but we have not changed the total value of the log because

$-2 + 1 =$ _____

- - - - - - - - - - - - - - - - - -

-1

18. $\bar{2} + 1.9085$ can be divided by 2.

$$\frac{-2 + 1.9085}{2} = \bar{1} + .9543 = \bar{1}.9543$$

Now all we have to do is find the antilog $\bar{1}.9543$ to get our answer for log $\sqrt[2]{.81}$.

antilog $\bar{1}.9543$ = _____

- - - - - - - - - - - - - - - - - -

.9

19. If we have a number with a negative characteristic and a positive mantissa, we must change the ________________ before it can be divided.

- - - - - - - - - - - - - - - - - -

characteristic

20. In changing the characteristic, the total algebraic value must remain the same but be evenly divisible by the divisor (denominator); for instance,

$\bar{7}.6654 \div 3 = (\bar{9} + 2.6654) \div 3 =$ ___________

<u>Note</u>: The characteristic $\bar{9}$ is now evenly divisible by 3.

- - - - - - - - - - - - - - - - - -

$\bar{3} + .8884$ or $\bar{3}.8884$

21. Perform these operations:

$\bar{1}.9085 \div 2 =$ _____ $+ 1.9085 \div 2$

$\bar{2}.4314 \div 3 =$ _____ $+ 1.4314 \div 3$

- - - - - - - - - - - - - - - - - -

$\bar{2}$

$\bar{3}$

22. Perform these operations:

$\bar{2}.8062 \div 3 =$ _____ + _________ $\div 3$

$\bar{9}.5051 \div 5 =$ _____ + _________ $\div 5$

- - - - - - - - - - - - - - - - - - -

$\bar{3} + 1.8062$

$\overline{10} + 1.5051$

23. Solve this problem:

$\log \sqrt[3]{.027} = (\log .027) \div$ _____ $= \bar{2}.4314 \div$ _____

$=$ (_____ $+ 1.4314) +$ _____ $=$ _____ $+ .4771 =$ ________

antilog __________ = _______

- - - - - - - - - - - - - - - - - - -

$\log \sqrt[3]{.027} = (\log .027) \div 3 = \bar{2}.4314 \div 3$

$= (\bar{3} + 1.4314) \div 3 = \bar{1} + .4771 = \bar{1}.4771$

antilog $\bar{1}.4771 = 0.3$

24. Solve this problem:

$\log \sqrt[3]{.064} = (\log .064) \div 3 =$ _______ $\div$ ______

$=$ (______ + ___________) $\div$ _____ $=$ _____ + __________

$=$ _______

antilog ____________ = ______

- - - - - - - - - - - - - - - - - - -

$\log \sqrt[3]{.064} = (\log .064) \div 3 = \bar{2}.8062 + 3$

$= (\bar{3} + 1.8062) \div 3 = \bar{1} + .6021$

$= \bar{1}.6021$

$\bar{1}.6021 = 0.4$

25. Solve this problem:

$\log \sqrt[5]{.032} =$ _____

- - - - - - - - - - - - - - - - - - -

$(\log .032) \div 5 = \bar{2}.5051 \div 5 = (\bar{5} + 3.5051) \div 5$

$= \bar{1} + .7010 = \bar{1}.7010$, antilog $\bar{1}.7010 = .502$

26. Solve this problem:

$\log \sqrt[5]{.007645} =$ ________

- - - - - - - - - - - - - - - - - - -

$(\log .007645) \div 5 = \bar{3}.8834 \div 5 = (\bar{5} + 2.8834)$

$= \bar{1}.5767$

antilog $\bar{1}.5767 = .3773$

27. There is another way of writing equations such as

$\log \sqrt[2]{10,000} = (\log 10,000) \div 2$

which can be written as

$\log \sqrt[2]{10,000} = \frac{1}{2} \log 10,000$

This is read "one half the log of 10,000."

Write $\log \sqrt[3]{1,000,000} = (\log 1,000,000) \div 3$ in another form.

_____ log 1,000,000

- - - - - - - - - - - - - - - - - - - -

$\frac{1}{3}$

28. Write $\log \sqrt[5]{2856} = (\log 2856) \div 5$ another way:

____

- - - - - - - - - - - - - - - - - - - -

$\frac{1}{5} \log 2856$

29. Log $\sqrt[7]{7865}$ = (log 7865) ÷ 7 = $\frac{1}{7}$ log 7865. $\frac{1}{7}$ log 7865 appears as a multiplication. Regardless of the way it is presented, a division is required to get the solution. Therefore

$\frac{1}{7}$(3.8957) = 3.8957 ÷ 7 = ________

- - - - - - - - - - - - - - - - - - -

.5565

30. Likewise, log $\sqrt[3]{.027}$ = $\frac{1}{3}$ log .027 = $\frac{1}{3}(\bar{2}.4314)$ = $\bar{2}.4314$ ÷ 3 = $(\bar{3} + 1.4314)$ ÷ 3 = ________

- - - - - - - - - - - - - - - - - - -

$\bar{1}.4771$

31. log $\sqrt[5]{.007645}$ = $\frac{1}{5}$ log .007645 = ________ = ______ ÷ ______ = (______ + ________) ÷ ______ = ________

- - - - - - - - - - - - - - - - - - -

log $\sqrt[5]{.007645}$ = $\frac{1}{5}$ log .007645 = $\frac{1}{5}(\bar{3}.8834)$ = $\bar{3}.8834$

÷ 5 = $(\bar{5} + 2.8834)$ ÷ 5 = $\bar{1}.5767$

32. We are now coming to problems with complex terms such as the following:

$$\log \frac{1}{\sqrt[7]{7.865}}$$

Let's proceed step by step. We may write this expression as

$$\log 1 - \log \sqrt[7]{7.865} = \log 1 - (\log 7.866) \div 7$$

$$= 0 - (0.8957 \div 7)$$

$$= 0 - .1279$$

$$= -0.1279$$

What have we here? A negative mantissa? But we said that we cannot use negative mantissas, and they don't appear on our log table. If you feel strong enough, figure this one out by yourself.

33. What can we do? Obviously, we must change our mantissa to make it positive. Examine the following equation:

$$-0.1279 = -1 + .8721 = \bar{1}.8721$$

We have added a +1 to the mantissa and a -1 to the characteristic which changed the form of the mathematical expression but not the value. Now find the antilog $\bar{1}.8721$.

antilog $\bar{1}.8721$ = __________

- - - - - - - - - - - - - - - - - -

.7448

34. $$\log \frac{1}{\sqrt{712}} = \log 1 - \log \sqrt{712}$$

$$= 0 - (2.8525 \div 2)$$

$$= \underline{\hspace{3cm}}$$

- - - - - - - - - - - - - - - - - -

-(1.4262)

35. Now we end up with a negative characteristic as well as a negative mantissa. Let's try to change the form of the expression but not its value.

-(1.4262) = ____________

- - - - - - - - - - - - - - - - - -

-2 + .5738 or $\bar{2}$.5738. Remember, all we did was add a +1 to the mantissa and a -1 to the characteristic, leaving the value the same but resulting in a positive mantissa.

36. Now find antilog $\bar{2}$.5738.

antilog $\bar{2}$.5738 = ____________

- - - - - - - - - - - - - - - - - -

.03748

37. In problems involving negative mantissas change them into a form in which you will have a negative characteristic and a positive mantissa. Remember, the values must remain the same.

$$\frac{1}{\sqrt[3]{5748}} = \log \frac{1}{\sqrt[3]{5748}} = \log 1 - \log \sqrt[3]{5748}$$

$$= 0 - (3.7595 \div 3)$$

$$= 0 - (1.2532)$$

$$= \bar{2}.7468$$

antilog $\bar{2}$.7468 = ____________

- - - - - - - - - - - - - - - - - -

.05582 or .05583

38. Perform the following operations

(a) $\frac{1}{\sqrt[5]{5.798}}$ = ____________

(b) $\frac{1}{\sqrt[4]{695}}$ = ____________

(c) $\frac{1}{\sqrt[6]{8756}}$ = ____________

(d) $\frac{1}{\sqrt[7]{67.98}}$ = ____________

- - - - - - - - - - - - - - - - - - -

(a) .7035

(b) .1947

(c) .2202 or .2203

(d) .547

39. In finding the log in this problem, and again in determining the antilog, we proceed step by step:

$$\begin{aligned}\log \frac{1}{\sqrt[5]{.5798}} &= \log 1 - \log \sqrt[5]{.5798} \\ &= 0 - \bar{1}.7633 \div 5 \\ &= 0 - [(5 + 4.7633) \div 5] \\ &= 0 - [\bar{1} + .95266] \\ &= 0 + 1 - .95266\end{aligned}$$

We wind up with a positive characteristic and a negative mantissa, which is quite unusual but really easier to resolve than the reverse condition.

$1 - .95266 = .04734$

This is a positive mantissa and easy to work with

antilog .04734 = __________

- - - - - - - - - - - - - - - - - -

2.974

40. $\log \frac{1}{\sqrt[3]{.0574}} = \log 1 - \log \sqrt[3]{.0574}$

$= 0 - [\bar{2}.7589 \div 3]$

$= 0 - [(3 + 1.7589) \div 3]$

$= 0 - [1 + .5863]$

$= 0 + 1 - .5863$

$=$ ____________

antilog ____________ = ____________

- - - - - - - - - - - - - - - - - - - -

.4137

.4137 = 2.593

41. $\log \frac{1}{\sqrt{.000574}} = \log 1 - \log \sqrt{.000574}$

$= 0 - (\bar{4}.7589 \div 2)$

$= 0 - (\bar{2}.3794\not{5})$

$= 2 - .3794$

$=$ ____________

antilog ____________ = ____________

- - - - - - - - - - - - - - - - - - - -

1.6206

1.6206 = 41.74

42. Using logs, find the answers to the following problems:

(a) $\dfrac{1}{\sqrt[4]{.000695}}$ = ____________

(b) $\dfrac{1}{\sqrt[3]{.005798}}$ = ____________

(c) $\dfrac{1}{\sqrt[7]{.00006798}}$ = ____________

(d) $\dfrac{1}{\sqrt[3]{.0436}}$ = ____________

- - - - - - - - - - - - - - - - - -

(a) 6.158 or 6.159

(b) 5.566 or 5.567

(c) 3.939

(d) 2.841 or 2.842

43. Let's try a slightly different problem:

$$\log \frac{7}{\sqrt[3]{5895}} = \log 7.5 - \log \sqrt[3]{5895}$$

$$= .8775 - (3.7705 \div 3)$$

$$= .8775 - 1.2568$$

$$= -1 + 1.8775 - 1.2568$$

$$= -1 + .6207$$

$$= \bar{1}.6207$$

$$\text{antilog } \bar{1}.6207 = \underline{\qquad\qquad}$$

- - - - - - - - - - - - - - - - - - -

.4176

Again, we have changed the form of a log but not its net value. We have added (-1 +1) to log .8775 and performed the subtraction without involved calculations.

44. Using logs, find the answers to the following problems:

(a) $\log \frac{57.9}{\sqrt[5]{8539}} = \underline{\qquad\qquad}$

(b) $\log \dfrac{3.5}{\sqrt[4]{6745}}$ = ______________

(c) $\log \dfrac{1.15}{\sqrt{3157}}$ = ______________

- - - - - - - - - - - - - - - - - - -

(a) $\log 57.9 - \log \sqrt[5]{8539}$

$= 1.7627 - (3.9314 \div 5)$

$= 1.7627 - .7863$

$= .9764$

antilog $.9764 = 9.472$ or 9.473

(b) $\log 3.5 - \log \sqrt[4]{6745}$

$= 1.5441 - (3.8290 \div 4)$

$= .5441 - .9573$

$= -1 + 1.5441 - .9573$

$= -1 + .5868$

$= \bar{1}.5868$

antilog $\bar{1}.5868 = .3862$

(c) $\log 1.15 - \log \sqrt{3157}$

$= .0607 - (3.4993 \div 2)$

$= .0607 - 1.7497$

$= -2 + 2.0607 - 1.7497$

$= -2 + .3110$

$= \bar{2}.3110$

antilog $\bar{2}.3110 = .02046$ or $.02047$

45. The following problem should not present any difficulty. Just follow the rules.

$$\log \frac{\sqrt[3]{7896}}{\sqrt[5]{9645}} = \left[(\log 7896) - 3\right] - \left[(\log 9645) - 5\right]$$
$$= \underline{\hspace{4cm}}$$

Complete the computations.

- - - - - - - - - - - - - - - - - - - -

3.178

46. Perform the following operations:

(a) $\log \dfrac{\sqrt[7]{8431}}{\sqrt[5]{7893}}$ = ________________

(b) $\log \dfrac{\sqrt[6]{53.75}}{\sqrt{.00135}}$ = ________________

(c) $\log \dfrac{\sqrt{3914}}{\sqrt[3]{5.116}}$ = ________________

(d) $\log \dfrac{\sqrt[7]{3575}}{\sqrt{.8571}}$ = ________________

- - - - - - - - - - - - - - - - - -

(a) .6044

(b) 52.87

(c) 36.31

(d) 3.475 or 3.476

47. Common logs are a form of exponential numbers in which the __________ (base, exponent) is always __________ (1, 10, 100, 1000) and is left out.

- - - - - - - - - - - - - - - - - - -

base

10

48. Multiplication is reduced to __________ of logs.

- - - - - - - - - - - - - - - - - - -

addition

49. Division is reduced to __________ of logs.

- - - - - - - - - - - - - - - - - - -

subtraction

50. Raising to powers is reduced to __________ of logs.

- - - - - - - - - - - - - - - - - - -

multiplication

51. Extracting roots is reduced to __________ of logs.

- - - - - - - - - - - - - - - - - - -

division

SELF-TEST

1. $\sqrt[3]{9867}$ = ______________

2. $\sqrt[6]{7589}$ = ______________

3. $\sqrt[7]{3.559}$ = ______________

4. $\sqrt[3]{.7398}$ = ______________

5. $\sqrt[4]{.007695}$ = ______________

6. $\dfrac{1}{\sqrt[7]{5964}}$ = ______________

7. $\dfrac{1}{\sqrt[5]{12.65}}$ = ______________

8. $\dfrac{1}{\sqrt[9]{.07865}}$ = ______________

9. $\dfrac{1}{\sqrt[8]{5.85 \times 10^{16}}}$ = ______________

10. $\dfrac{5.4}{\sqrt[9]{7865}}$ = ______________

11. $\dfrac{351.7}{\sqrt[6]{381.6}}$ = ______________

12. $\dfrac{17.5}{\sqrt[5]{.00375}}$ = ______________

13. $\dfrac{\sqrt[3]{5813}}{\sqrt{6835}}$ = ______________

14. $\dfrac{\sqrt[5]{9763}}{\sqrt[3]{4473}}$ = ______________

15. $\dfrac{89.75}{\sqrt[5]{92150}}$ = ______________

Answers

1. (log 9867) ÷ 3 = 3.9942 ÷ 3
 = 1.3314
 antilog 1.3314 = 21.45

2. (log 7589) ÷ 6 = 3.8802 ÷ 6
 = .6467
 antilog .6467 = 4.433

3. (log 3.559) ÷ 7 = .5513 ÷ 7
 = .0788
 antilog .0788 = 1.199

4. (log .7398) ÷ 3 = $\bar{1}.8686 \div 3$
 = $(\bar{3} + 2.8686) \div 3$
 = $\bar{1}.9562$
 antilog $\bar{1}.9562$ = .9050

5. (log .007695) ÷ 4 = $\bar{3}.8862 \div 4$
 = $(\bar{4} + 1.8862) \div 4$
 = $\bar{1}.4716$
 antilog $\bar{1}.4716$ = .2962

6. $\log 1 - \left[(\log 5964) \div 7\right] = 0 - (3.7755 \div 7)$

$= 0 - .5394$

$= \overline{1}.4606$

antilog $\overline{1}.4606 = .2888$

7. $\log 1 - \left[(\log 12.65) \div 5\right] = 0 - (1.1021 \div 5)$

$= 0 - .2204$

$= \overline{1}.7796$

antilog $\overline{1}.7796 = .602$

8. $\log 1 - \left[(\log .07865) \div 9\right] = 0 - (\overline{2}.8957 \div 9)$

$= 0 - \left[(\overline{9} + 7.8957) \div 9\right]$

$= 0 - \overline{1}.8773$

$= .1227$

antilog $.1227 = 1.326$ or 1.327

9. $\log 1 - \frac{1}{8} \log (5.85 \times 10^{16})$ or

$\log 1 - \left[\log (5.85 \times 10^{16}) \div 8\right] = 0 - (16.7672 — 8)$

$= 0 - 2.0959$

$= \overline{3}.9041$

antilog $\overline{3}.9041 = .008019$

10. $\log 5.4 - \left[(\log 7865) \div 9\right] = .7324 - (3.8957 \div 9)$

$= .7324 - .4329$

$= .2995$

antilog $.2995 = 1.993$

11. $\log 351.7 - \left[(\log 381.6) \div 6\right] = 2.5462 - (2.5816 \div 6)$

$= 2.5462 - .4303$

$= 2.1159$

antilog 2.1159 = 130.6

12. $\log 17.5 - \left[(\log .00375) \div 5\right] = 1.2430 - (\bar{3}.5740 \div 5)$

$= 1.2430 - \left[(\bar{5} + 2.5740) \div 5\right]$

$= 1.2430 - \bar{1}.5148$

$= 1.7282$

antilog 1.7282 = 53.49

13. $\left[(\log 5813) \div 3\right] - \left[(\log 6835) \div 2\right] = (3.7644 \div 3)$

$- (3.8347 \div 2)$

$= 1.2548 - 1.9174$

$= (\bar{1} + 2.2548)$

$- 1.9174$

$= \bar{1}.3374$

antilog $\bar{1}.3374$ = .2174 or .2175

14. $\left[(\log 9763) \div 5\right] - \left[(\log 4473) \div 3\right] = (3.9895 \div 5)$

$- (3.6506 \div 3)$

$= .7979 - 1.2169$

$= \bar{1}.5810$

antilog $\bar{1}.5810$ = .3811

15. $\left[(\log 89.75) \div 2\right] - \left[(\log 91250) - 5\right] = (1.9530 \div 2)$

$- (4.9645 \div 5)$

$= .8765 - .9929$

$= \bar{1}.8836$

antilog $\bar{1}.8836$ = .7648 or .7649

UNIT FOURTEEN

One More Way to Write Negative Logarithms

1. You have learned to write negative logs in three ways; for instance, $\overline{7}.9571$ may also be expressed as _____ or _____.

- - - - - - - - - - - - - - - - - -

-7 + ,9571 or .9571 -7

2. A negative log may be written in any convenient form, but, remember, the net value of the expression must remain the same. Now complete this simple equation:

-7 = ____________ -10

- - - - - - - - - - - - - - - - - -

3

3. Complete this equation:

.9571 -7 = _______.9571 -10

- - - - - - - - - - - - - - - - - -

3

4. $\overline{7}.9571$ or (-7 + .9571) or (.9571 -7) may also be written as (3.9571 -10). Express the following negative logs in the new form:

$\overline{2}.3979$ = ________		$\overline{4}.9047$ = ________	
$\overline{1}.3222$ = ________		$\overline{6}.9031$ = ________	
$\overline{1}.9047$ = ________		$\overline{5}.9222$ = ________	
$\overline{2}.4771$ = ________		$\overline{8}.0016$ = ________	

8.3979 - 10	6.9047 - 10
9.3222 - 10	4.9031 - 10
9.9047 - 10	5.9222 - 10
8.4771 - 10	2.0016 - 10

5. Add each group of four quantities in frame 4.

__________ - __________

__________ - __________

- - - - - - - - - - - - - - - - - - -

36.1019 - 40

19.7316 - 40

6. Changing the form of these two quantities, but not their net values, you may write them as

___.1019 and ___. __________

- - - - - - - - - - - - - - - - - - -

$\bar{4}.1019$ or $-4 + .1019$ or $.1019 - 4$

$\overline{21}.7316$ or $-21 + .7316$ or $.7316 - 21$

7. Complete the second column and add both.

$\bar{5}.7641$ = __________ -

2.3786 = __________

$\overline{14}.7789$ = __________ -20

1.1789 = __________

3.4137 = __________

$\bar{8}.6449$ = ____________ -

______ = ____________ -

- - - - - - - - - - - - - - - - -

	5.7641	-10
	2.3786	
	6.7789	-20
	1.1789	
	3.4137	
______	2.6449	-10
18.1591 =	22.1591	-40

8. Using logs, solve the following problem:

$765.3 \times 8.415 \times .3756 \times 91.73 \times (7.155 \times 10^{-15})$

$\times (5.81 \times 10^{13}) \times .003741 =$ ______________

- - - - - - - - - - - - - - - - -

log 765.3 =	2.8839	
log 8.415 =	.9251	
log .3756 =	9.5747	-10

log 91.73	=	1.9625	
log (7.155×10^{-15})	=	5.8546 -20	
log (5.81×10^{13})	=	13.7642	
log .003741	=	7.5730 -10	
		42.5380 -40	= 2.5380
antilog 2.5380	=	345.1 or 345.2	

9. As long as the net value remains the same, the number may be changed to facilitate a calculation. The following logs may also be written:

$\bar{3}.5173$	$\overline{23}.5173$
______ + .______	______ + .______
.______ - ______	.______ - ______
___.___ - ______	___.___ - ______

- - - - - - - - - - - - - - - - - - -

-3 + .5173	-23 + .5173
.5173 - 3	.5173 - 23
7.5173 - 10	7.5173 - 30

SELF-TEST

Change the numbers below using the 10 - 10 method (or multiples of 10 - 10).

1. $\bar{3}.1748$ = ______________

2. $\overline{15}.3897$ = ______________

3. $\bar{4}.1791$ = ______________

4. $\overline{27}.1891$ = ____________________

5. $\overline{33}.9753$ = ____________________

6. $\overline{1}.7783$ = ____________________

7. $\overline{5}.1793$ = ____________________

8. $\overline{17}.0371$ = ____________________

9. $\overline{6}.8579$ = ____________________

10. .7789 -3 = ____________________

11. .6655 -17 = ____________________

12. .9978 -36 = ____________________

13. .1718 -1 = ____________________

14. .5639 -9 = ____________________

15. .6773 -25 = ____________________

Answers

1. 7.1748 -10

2. 5.3897 -20

3. 6.1791 -10

4. 3.1891 -30

5. 7.9753 -40

6. 9.7783 -10

7. 5.1793 -10

8. 3.0371 -20

9. 4.8579 -10

10. 7.7789 -10

11. 3.6655 -20

12. 4.9978 -40

13. 9.1718 -10

14. 1.5639 -10

15. 5.6773 -30

UNIT FIFTEEN

Cologs and Involved Problems

1. $\log \frac{1}{3} = \log 1 - \log 3$

   $= 0 - .4771$

   $\log \frac{1}{5} = \log 1 - \log 5$

   $= 0 - .6990$

2. In both cases the results are negative mantissas which can be changed into positive mantissas by adding (10 - 10), keeping the net value the same.

   $-.4771 + (10 - 10) = 9.5229 - 10$

   $-.6990 + (10 - 10) =$ ______ - ____

9.3010   10

This method comes in very handy when working calculations with many multiplications and divisions. Instead of subtracting a log, we add its so-called colog. The subtraction is actually done by finding the colog; for instance,

$\text{colog } \mathbf{5} = \log \frac{1}{5} = \log 1 - \log 5 = \underline{0 - .6990}\ -.6990$

$= 9.3010 - 10$ subtraction

3. Find

(a) colog 3.591 $= \log \dfrac{1}{3.591} =$ ________

(b) colog 35.91 = ____________

(c) colog 7653 = ____________

- - - - - - - - - - - - - - - - - - - -

(a) log 1 - log 3.591

0 -.5552 = - .5552

= 9.4448 - 10

(b) $\log \dfrac{1}{35.91}$ = log 1 - log 35.91

= 0 - 1.5552 = -1.5552

= 8.4448 -10

(c) $\log \dfrac{1}{7653}$ = log 1 - log 7653

= 0 -3.8837 = -3.8837

= 6.1163 -10

4. Find the cologs for the following numbers:

colog .7935 = .1004

(a) colog 8.315 = _______

(b) colog .08836 = _______

(c) colog .005637 = _______

(a) 9.0801 -10

(b) 1.0537

(c) 2.2490

5. In your calculations you applied the general formula which states that the colog of number N equals the log of the reciprocal $\frac{1}{N}$ or, expressed differently,

$$\text{colog } N = \log \frac{1}{N} = \log 1 - \log N = -\log N$$

Take a good look at -log N, using two examples:

colog 35.91 = 1.5552 = -1 - .5552

colog .571 = $\overline{1}.7566$ = 1 - .7566

In the first example the characteristic is negative and in the second it is positive, but in both examples the mantissa is negative. If this appears inconvenient for further calculations, the mantissa can be made positive without changing the net value of the colog.

-1.5552 = 9.4448 -10

$-\overline{1}.7566$ = .2434

6. Change the following numbers so that the mantissas become positive:

- 5.3775 = ____________

- 4.7663 = ____________

- 15.8115 = ____________

- 7.3571 = ____________

- $\overline{5}.3775$ = ____________

- $\overline{4}.7663$ = ____________

- $\overline{15}.8115$ = ____________

- $\overline{7}.3571$ = ____________

- - - - - - - - - - - - - - - - - -

4.6225 -10

5.2337 -10

4.1885 -20

2.6419 -10

4.6225

3.2337

14.1885

6.6429

7. Suppose you had to perform the following calculations:

$$\frac{369.7 \times 1785 \times 5135}{571 \times 77.5 \times 1835}$$

We can change this combination to a multiplication:

$$369.7 \times 1785 \times 5135 \times \frac{1}{571} \times \frac{1}{77.5} \times \frac{1}{1835}$$

By using logs it is written

log 369.7 + log 1785 + log 5135 + colog 571
+ colog 77.5 + colog 1835 = ____________________

- - - - - - - - - - - - - - - - - -

log 369.7	=	2.5678		
log 1785	=	3.2516		
log 5135	=	3.7105		
colog 571	=	7.2434	-10	
colog 77.5	=	8.1107	-10	
colog 1835	=	6.7363	-10	
antilog	=	31.6203	-30	= 41.72

8. Changing the fraction to a multiplication was done to show the logical development of the method used. In actual calculation all that is necessary is to find the logs for all the numbers above the line and the cologs for all the numbers below the line. Perform the calculations:

$$\frac{1785 \times 13.64 \times .0765 \times .895 \times .0031}{587.5 \times 17.6 \times .3115 \times 00315} = \underline{\qquad\qquad}$$

- - - - - - - - - - - - - - - - - -

log 1785	=	3.2516	
log 13.64	=	1.1348	
log .0765	=	8.8837	- 10
log .895	=	9.9518	- 10
log .0031	=	7.4914	- 10
colog 587.5	=	7.2310	- 10
colog 17.6	=	8.7545	- 10
colog .3115	=	.5065	
colog .00315	=	2.5017	
		49.7070	- 50 = $\bar{1}.7070$
antilog $\bar{1}.7070$	=	.5093 or .5094	

9. Let's try this calculation:

$$\sqrt[3]{.5789} =$$

$$\log \sqrt[3]{.5789} = (\log .5789) \div 3$$
$$= \bar{1}.7626 \div 3$$
$$= (29.7626 - 30) \div 3$$
$$= 9.9209 - 10$$
$$= \bar{1}.9209$$
$$\text{antilog } \bar{1}.9209 = .8335 \text{ or } .8336$$

We added (30 - 30) to the log to make the negative characteristic divisible by 3 and to obtain a -10 in the result of the division. Perform the following calculations:

(a) $\sqrt[5]{.1789}$

(b) $\sqrt[4]{.01789}$

(c) $\sqrt[6]{.001789}$

(d) $\sqrt[13]{.001789}$

- - - - - - - - - - - - - - - - - - - -

(a) $\log \sqrt[5]{.1789} = (\log .1789) \div 5$

$= \bar{1}.2526 \div 5$

$= (49.2526 - 50) \div 5$

$9.8505 - 10$

$= \bar{1}.8508$

$\text{antilog } \bar{1}.8505 = .7088$

(b) $\log \sqrt[4]{.01789} = (\log .01789) \div 4$

$= \bar{2}.2526 \div 4$

$= (38.2526 - 40) \div 4$

$= 9.5632 - 10$

$= \bar{1}.5632$

antilog $\bar{1}.5632 = .3657$ or $.3658$

(c) $\log \sqrt[6]{.001789} = (\log .001789) \div 6$

$= \bar{3}.2526 \div 6$

$= (57.2526 - 60) \div 6$

$= 9.5421 - 10$

$= \bar{1}.5421$

antilog $\bar{1}.5421 = .3484$

(d) $\log \sqrt[13]{.001789} = (\log .001789) \div 13$

$= \bar{3}.2526 \div 13$

$= (127.2526 \div 130) \div 13$

$= 9.7887 - 10$

$= \bar{1}.7887$

antilog $\bar{1}.7887 = .6147$

Each line of these problems (and those that follow) could be solved by the method used in Unit 13, but this new method may be more convenient for calculations with mixed numbers. If you feel strong enough, you could omit some of the steps on paper and do them in your head.

10. Perform the following calculations:

$$(793.5)^5 \times \frac{(369.7)^3 \times .1785 \times \sqrt[3]{.5813} \times .1775}{\sqrt{657} \times (367)^2 \times .1975}$$

= ____________________

- - - - - - - - - - - - - - - - - -

5 log 793.5	=	2.8996 x 5	=	14.4980
3 log 369.7	=	2.5678 x 3	=	7.7034
log .1785	=	$\bar{1}.2516$	=	9.2516 -10
log $\sqrt[3]{.5813}$	=	(29.7644 -30) ÷ 3	=	9.9215 -10
log .1775	=	$\bar{1}.2492$	=	9.2492 -10
colog $\sqrt{675}$	=	(17.1707 -20) ÷ 2	=	8.5854 -10
2 colog 367	=	2(7.4353 -10)	=	14.8706 -20
colog .1975	=			.7044
				74.7841 -60
			=	14.7841

antilog 14.7841 = 6.083×10^{14}

11. As we have seen mathematical quantities can be expressed in various forms; for instance,

$\log (84)^2 = 2 \log 84$

$\sqrt[5]{975} = (975)^{\frac{1}{5}}$ $\qquad \log \sqrt[5]{975} = \frac{1}{5} \log 975$

$75 \times 85 = (75)\ (85)$ $\qquad \log (75 \times 85) = \log 75 + \log 85$

Solve the problem below by using the above form of expressions:

$$\frac{.008753 \times .7983 \times (884.1)^3 \times \sqrt[3]{779.5}}{(339.1)^{\frac{1}{5}} \times \sqrt[7]{.553}} = \underline{\qquad\qquad}$$

- - - - - - - - - - - - - - - - - - -

$$\log \frac{(.008753)\ (.7983)\ (884.1)^3\ (779.5)^{\frac{1}{3}}}{(339.1)^{\frac{1}{5}}\ (.553)^{\frac{1}{7}}}$$

log .008753	=	$\bar{3}.9421$	= 7.9421 -10
log .7983	=	$\bar{1}.9022$	= 9.9022 -10
3 log 884.1	=	2.9465 x 3	= 8.8395
$\frac{1}{3}$ log 779.5	=	2.8918 ÷ 3	= .9639
$\frac{1}{5}$ colog 339.1	=	$\frac{1}{5}$ ($\bar{3}$.4697) = $\frac{47.4697-50}{5}$	= 9.4939 -10
$\frac{1}{7}$ colog .553	=	$\frac{1}{7}$ (.2573) =	.0368
			=37.1784 -30
			= 7.1784

antilog 7.1784 = 1.508×10^7

12. You may encounter a problem similar to the one below:

$\sqrt[3]{-27}$

Of course, you don't need logs to find the root. Notice that the root must be a negative number: $-3^3 = -3 \times -3 \times -3 = -27$. Using logs, work the problem as follows:

$$\log \sqrt[3]{-27} = \frac{1}{3} \log -27 = \frac{1}{3}(1.4314)n$$

$$= .4771n$$

$$\text{antilog } .4771n = -3$$

The n after the log has only one purpose: to indicate that the final result is negative.

Now find the following roots:

(a) $\sqrt[5]{-8971}$

(b) $\sqrt[7]{-.8511}$

(c) $\sqrt[3]{-4.711}$

(d) $\sqrt[9]{-.00567}$

- - - - - - - - - - - - - - - - - -

(a) $\frac{1}{5} \log -8971 = \frac{1}{5}(3.9528)n$

$= .7906n$

antilog $.7906n = -6.174$

(b) $\frac{1}{7} \log -.8511 = \frac{1}{7}(\bar{1}.9300)n$

$= \frac{69.9300 - 70}{7}n$

$= (9.9900 - 10)n$

$= \bar{1}.9900n$

antilog $\bar{1}.9900n = -.9770$

(c) $\frac{1}{3} \log -4.711 = \frac{1}{3}(.6731)n$

$= .2244n$

antilog $.2244n = -1.676$ or -1.677

(d) $\frac{1}{9} \log -.00567 = \frac{1}{9}(\bar{3}.7536)n$

$= \frac{87.7536 - 90}{9}n$

$= \bar{1}.7504n$

antilog $\bar{1}.7504n = -.5629$

SELF-TEST

1. colog 7.96 = ________
2. colog 796 = ________
3. colog 175.8 = ________
4. colog 17.58 = ________
5. colog .1758 = ________
6. colog 10^{6} = ________
7. colog 10^{-6} = ________
8. colog .001 = ________
9. colog .023 = ________
10. colog .0023 = ________

Using logs and cologs, perform the following calculations:

11. $\dfrac{(375.4)\ (57.79)}{(3.896)\ (16.57)}$ = ________

12. $\dfrac{(389)^{2}\ (786.9)^{3}\ (598)}{\sqrt[2]{(76.5)^3}\ \sqrt[5]{381.5}}$ = ________

13. $\dfrac{(521)^{\frac{1}{5}}\ (76.3)^{\frac{1}{2}}\ (735)}{(35.1)^{2}\ (983)^{\frac{1}{2}}\ (991)^{3}}$ = ________

14. $$\frac{(713.5)\,(853.6)^{3}\,(581.3)^{\frac{3}{2}}}{(76.5)^{\frac{1}{2}}\,(83)^{2}\,(971)^{2}} = \underline{\hspace{4cm}}$$

15. $$\frac{(3.13 \times 10^{3})\,(5.16 \times 10^{-3})\,(619.5)^{\frac{1}{3}}}{(93.5)^{2}\,(714.9)^{\frac{1}{5}}\,(17.1)^{2}} = \underline{\hspace{4cm}}$$

Answers

1. $\bar{1}.0991$

2. $\bar{3}.0991$

3. $\bar{3}.7550$

4. $\bar{2}.7550$

5. .7550

6. $\bar{6}.0000$

7. 6.0000

8. colog 10^{-3} = 3.0000

9. 1.6383

10. 2.6383

11.
$$\begin{array}{lllll}
\log 375.4 & = & & & 2.5745 \\
\log 57.79 & = & & & 1.7619 \\
\text{colog } 3.896 & = & \bar{1}.4094 & = & 9.4094 - 10 \\
\text{colog } 16.57 & = & \bar{2}.7807 & = & \underline{8.7807 - 10} \\
 & & & & 22.5265 - 20 = 2.5265
\end{array}$$

antilog 2.5265 = 336.1 or 336.2

12. 2 log 389 = 2.5899 x 2 = 5.1798

3 log 786.9 = 2.8959 x 3 = 8.6877

log 598 = 2.7767

$\frac{3}{2}$ colog 76.5 = $\overline{2}.1163 \times \frac{3}{2}$ = 7.1745 -10

$\frac{1}{5}$ colog 381.5 = $\overline{3}.4185 \div 5$ = 9.4837 -10

33.3024 -20 = 13.3024

antilog 13.3024 = 2.006×10^{13} or 2.007×10^{13}

13. $\frac{1}{5}$ log 521 = $2.7168 \div 5$ = .5434

$\frac{1}{2}$ log 76.3 = $1.8825 \div 2$ = .9412

log 735 = 2.8663

2 colog 35.1 = $\overline{2}.4547 \times 2$ = 6.9094 -10

$\frac{1}{2}$ colog 983 = $\overline{3}.0074 \div 2$ = 8.5037 -10

3 colog 991 = $\overline{3}.0039 \times 3$ = 1.0117 -10

20.7757 -30 = $\overline{10}.7757$

antilog $\overline{10}.7757$ = 5.967×10^{10}

14. log 713.5 = 2.8534

3 log 853.6 = 2.9312 x 3 = 8.7936

$\frac{3}{2}$ log 581.3 = $2.7644 \times \frac{3}{2}$ = 4.1466

$\frac{1}{2}$ colog 76.5 = $\overline{2}.1163 \div \frac{1}{2}$ = 9.0582 -10

2 colog 83 = $\overline{2}.0809 \times 2$ = 6.1618 -10

$\frac{1}{2}$ colog 971 $= \bar{3}.0128 \div 2 = 8.5064 - 10$

$39.5200 - 30 = 9.5200$

antilog $9.5200 = 3.311 \times 10^{9}$ or 3.312×10^{9}

15. $\log(3.13 \times 10^{3}) = \; = .4955$

$\log(5.16 \times 10^{-3}) = \; = .7126$

$\frac{1}{3}\log 619.5 = 2.7921 \div 3 = .9307$

2 colog $93.5 = \bar{2}.0292 \times 2 = 6.0584 - 10$

$\frac{1}{5}$ colog $714.9 = \bar{3}.1458 \div 5 = 9.4292 - 10$

2 colog $17.1 = \bar{2}.7670 \times 2 = 7.5340 - 10$

$25.1604 - 30 = \bar{5}.1604$

antilog $\bar{5}.1604 = 1.447 \times 10^{-5}$

UNIT SIXTEEN

Natural Logarithms

1. So far we have worked with logs to the base 10. These common logs are usually the most convenient, but often scientists and engineers use what we call natural logs for their calculations.

   The rules for common logs are also valid for natural logs. When using logs,

   multiplication is reduced to ____________________ of logs,

   division is reduced to ________________________ of logs,

   raising to powers is reduced to ________________ of logs,

   extracting roots is reduced to _________________ of logs.

- - - - - - - - - - - - - - - - - -

   addition

   subtraction

   multiplication

   division

   If you change numbers into logs, perform a calculation, and change back, it does not make any difference what log system you use. The result will be the same within limits of accuracy.

2. Natural logarithm has as its base the number 2.71828.... In mathematical shorthand this number is simply written as e, and $\log_e$ (or ln) is the log to the base e or 2.71828.... The table of natural logarithms is shown on the card insert and on page 214.

Find $\log_e 2$ = _____ $\log_e 3$ = _____ $\log_e 5$ = _____ $\log_e 7$ = _____

- - - - - - - - - - - - - - - - - -

0.6931 1.0986 1.6094 1.9459

3. Finding the $\log_e$ of number is not greatly different than finding common logs. However, you have to be careful; $\log_e 3.43 = 1.2326$. The first digit, in this case 1, must be inserted when finding logs. It is shown only in the 0 column or when it changes to the next higher number.

$\log_e 3.04$ = ____________

$\log_e 2.72$ = ____________

$\log_e 2.75$ = ____________

$\log_e 2.71$ = ____________

- - - - - - - - - - - - - - - - - -

1.1119

1.0006

1.0116

.9969

4. Interpolation is done the same way as in common logs, using the columns under "mean differences."

$$\log_e 4.230 = 1.4422$$
$$+.005 \quad + \quad .0012$$
$$\log_e 4.235 = 1.4434$$

(a) $\log_e 4.553$ = ________

(b) $\log_e 2.767$ = ________

(c) $\log_e 8.338$ = ________

- - - - - - - - - - - - - - - - - -

(a) 1.5158

(b) 1.0177

(c) 2.1209

5. So far we have been working with the logs of numbers 1 to 10. What about numbers greater than 10, for instance, $\log_e 250$? Again we break the number down into $\log_e (2.5 \times 10^2)$

$$\log_e (2.5 \times 10^2) = \log_e 2.5 + \log_e 10^2$$

$$= \log_e 2.5 + (\log_e 10) \times 2$$

$$= \underline{\qquad\qquad} + 2.3026 \times 2$$

$$= \underline{\qquad\qquad} + 4.6052$$

$$= \underline{\qquad\qquad}$$

- - - - - - - - - - - - - - - - - -

0.9163

0.9163

5.5215

Let's repeat portions of our example in slow motion:

$$\log_e 250 = \log_e (2.5 \times 10^2)$$

$= \log_e 2.5 + \log_e 10^2$ — multiplication changes to addition

$= \log_e 2.5 + (\log_e 10) \times 2$ — raising to power changes to multiplication

These are the same rules we followed in working with common logs. Note, however, that the exponent 2 does not become the characteristic; instead ($\log_e 10$) must be treated as a separate $\log_e$, multiplied by 2, and then added to the $\log_e 2.5$.

6. Perform the following operations:

(a) $\log_e 345$ = ________________

(b) $\log_e 786$ = ________________

(c) $\log_e 1000$ = ________________

- - - - - - - - - - - - - - - - - - -

(a) $\log_e 345 = \log_e 3.45 + \log_e 10^2$

$= \log_e 3.45 + (\log_e 10) \times 2$ or $\log_e 10$

$= 1.2384 + 2.3026 \times 2$

$= 1.2394 + 4.6052$

$= 5.8436$

(b) $\log_e 786 = \log_e 7.86 + 2 \log_e 10$ (skipping one step)

$= 2.0618 + 4.6052$ (skipping another step)

$= 6.6670$

(c) $\log_e 1000 = \log_e 10^3 = 3 \log_e 10 = 6.9078$

7. Logs of 10^n can be found simply by consulting the one-line table under "Natural Logarithms of 10^{+n}" at the bottom of page 214.

Find the logs for

(a) 10,000 = ____________

(b) 10^6 = ____________

(c) 10^9 = ____________

(d) 10 = ____________

- - - - - - - - - - - - - - - - - - - -

(a) 9.2103

(b) 13.8155

(c) 20.7233

(d) 2.3026

What about $\log_e 10^{10}$ or $\log_e 10^{15}$? These can be found by multiplying 2.3026 by 10 and 15, respectively, or by adding $\log_e 10^9 + \log_e 10^1$ and $\log_e 10^9 + \log_e 10^6$, respectively.

8. Find the $\log_e$ of the following:

$\log_e 10^{10}$ = ____________

$\log_e 10^{15}$ = ____________

$\log_e 10^{30}$ = ____________

- - - - - - - - - - - - - - - - - - - -

23.0260 or 23.0259

34.5390 or 23.5388

69.0780 (If you chose to multiply $\log_e 10^9$ by three and add $\log_e 10^3$, your answer would be 69.0777. Other ways of adding would result in similar variations, insignificant for most purposes.)

9. Let's try a simple multiplication:

$\log_e (2.495 \times 3.791) = \log_e 2.495 + \log_e 3.791$

= (a) ________ + ________

= (b) ________

(c) antilog ________ = ________

- - - - - - - - - - - - - - - - - - -

(a) 0.9143 + 1.3327

(b) 2.2470

(c) 2.2470 = 9.459

10. Find $\log_e (3.457 \times 4.78)$.

$\log_e 3.457 + \log_e 4.78 = 1.2404 + 1.5644$

= 2.8048

antilog 2.8048 = ?

Let's take a close look at the table of natural logarithms which shows as the highest number log 10 = 2.3026. The above antilog 2.8048 is larger than that number, however. At the bottom of page 214 there is a one-line table entitled "Natural Logarithms of 10^{+n}." Antilog 2.8048 falls between two numbers as shown below:

ln 1 (10^1) ln 2 (10^2)

2,3026 4.6052

2.8048

Difference = 0.5022

→Find on large table

We have fitted our number into the right slot and will break it down into

antilog (2.3026 + 0.5022)

or in different format

antilog 2.3026 = 10^1 (from one-line table)

antilog 0.5022 = 1.652 (from large table)

antilog 2.8048 = 1.652 x 10^1 = 16.52

11. Find antilog 9.5376.

- - - - - - - - - - - - - - - - - - - -

antilog 9.2103 = 10^4

antilog 0.3273 = 1.387

antilog 9.5376 = 1.387 x 10^4

12. Operations that use natural logs are no different from those that use common logs. The only difficulty, which is a minor one, is in finding logs and antilogs on the tables. A little practice will be helpful.

Find antilogs

(a) antilog 13.6751 = ________________

(b) antilog 16.3191 = ________________

(c) antilog 20.8135 = ________________

(d) antilog 4.7813 = ________________

- - - - - - - - - - - - - - - - - - - -

(a) antilog 13.6751

antilog 11.5129	=	10^5
antilog 2.1622	=	8.690
antilog 13.6751	=	8.69×10^5

(b) antilog 16.3191

antilog 16.1181	=	10^7
antilog 0.2010	=	1.222
antilog 16.3191	=	1.222×10^7

(c) antilog 20.8135

antilog 20.7233	=	10^9
antilog 0.0902	=	1.094
antilog 20.8135	=	1.094×10^9

(d) antilog 4.7813

antilog 4.6052 $= 10^2$

antilog $\underline{0.1761}$ $= 1.192$

antilog 4.7813 $= 1.192 \times 10^2$

$= 119.2$

13. Up to this point we have dealt with numbers larger than 1. What about fractions? As in common logs, a number smaller than 1 will have a negative $\log_e$. Again we break the number down into two parts as shown below:

$$\log_e 0.00356 = \log_e (3.56 \times 10^{-3})$$
$$= \log_e 3.56 + \log_e 10^{-3}$$
$$= 1.2698 + \bar{7}.0922$$
$$= \bar{6}.3620$$

Log_e 3.56 was found on the large table and $\log_e 10^{-3}$ on the one-line table under "Natural Logarithms of 10^{-n}."

Find the following logs:

(a) log 0.07181 = ________________

(b) log 0.0004178 = ______________

(c) $\log 8.198 \times 10^{-15} =$ ______________

- - - - - - - - - - - - - - - - - -

(a) $\log_e 0.7181 = \log_e 7.181 + \log_e 10^{-2}$

$= 1.9714 + \bar{5}.3948$

$= \bar{3}.3662$

(b) $\log_e 0.0004178 = \log_e 4.178 + \log_e 10^{-4}$

$= 1.4298 + \overline{10}.7897$

$= \bar{8}.2195$

(c) $\log_e (8.198 \times 10^{-15}) = \log_e 8.198 + \log_e 10^{-15}$

$= 2.1039 + \overline{35}.4612$

$= \overline{33}.5651$

14. All numbers shown on the log tables are positive except for the one-line table of 10^{-n} in which only the whole numbers, the integers, are negative. Let's find antilog $\bar{8}.2195$.

Step 1. Fit the number in the right slot as shown below:

n3 (10^{-3})		n4 (10^{-4})
$\bar{7}.0922$	$\bar{8}.2195$	$\overline{10}.7897$

Difference 1.4298
find on large table

Notice that in finding an antilog you look for next lower number on the table. It appears that in the last frame the next higher number was used, but actually the number was lower in value because we are dealing with negative quantities; for instance, 10^{-1} dollars is 10 cents, but 10^{-2} dollars is only 1 cent. Therefore to find the difference between two numbers you subtract the smaller from the larger. In our example the difference is found as follows:

larger number	$\bar{8}.2195$
- smaller number	$-\overline{10}.7897$
difference	1.4298

Step 2. Break the number down into

antilog ($\overline{10}.7897$ + 1.4298)

or in a different format

antilog $\overline{10}.7897$ = 10^{-4} (from one-line table)

antilog 1.4298 = 4.178 (from large table)

antilog $\bar{8}.2195$ = 4.178×10^{-4} or 0.0004178

Now find the following antilogs:

(a) antilog $\overline{11}.5785$ = ______________

(b) antilog $\bar{6}.8791$ = ______________

(c) antilog $\overline{13}.4871$ = ______________

(d) antilog $\overline{17}.9785$ = ______________

- - - - - - - - - - - - - - - - - - -

(a) large number $\overline{11}.5785$ antilog $\overline{12}.4871 = 10^{-5}$

small number $\underline{\overline{12}.4871}$ antilog $\underline{1.0914} = 2.978$

difference 1.0914 antilog $\overline{11}.5785 = \underline{2.978 \times 10^{-5}}$

(b) large number $\overline{6}.8791$ antilog $\overline{7}.0922 = 10^{-3}$

small number $\underline{\overline{7}.0922}$ antilog $\underline{1.7869} = 5.971$

difference 1.7869 antilog $\overline{6}.8791 = \underline{5.971 \times 10^{-3}}$

(c) large number $\overline{13}.4871$ antilog $\overline{14}.1845 = 10^{-6}$

small number $\underline{\overline{14}.1845}$ antilog $\underline{2.3026} = 3.679$

difference 1.3026 antilog $\overline{13}.4871 = \underline{3.679 \times 10^{-6}}$

(d) large number $\overline{17}.9785$ antilog $\overline{17}.8819 = 10^{-7}$

small number $\underline{\overline{17}.8819}$ antilog $\underline{0.0966} = 1.101$ or 1.102

difference 0.0966 antilog $\overline{17}.9785 = \underline{1.101 \times 10^{-7}}$

or

$\underline{1.102 \times 10^{-7}}$

15. When finding antilogs on the one-line table, you must consider the negative integer and the positive decimal fraction as parts of one number; for instance: $\overline{17}.8819$ is lower than 17 .9785.

Find the antilogs

(a) antilog $\overline{10}.8765$ = ____________________

(b) antilog $\overline{12}.6761$ = ____________________

(c) antilog $\overline{14}.7356$ = ____________________

- - - - - - - - - - - - - - - - - -

(a) antilog $\overline{10}.7897$ = 10^{-4}

antilog $\underline{0.0868}$ = 1.091

antilog $\overline{10}.8765$ = $\underline{1.091 \times 10^{-4}}$

(b) antilog $\overline{12}.4871$ = 10^{-5}

antilog $\underline{0.1890}$ = 1.208

antilog $\overline{12}.6761$ = $\underline{1.208 \times 10^{-5}}$

(c) antilog $\overline{14}.1845$ = 10^{-6}

antilog $\underline{0.5511}$ = 1.735 or 1.736

antilog $\overline{14}.7356$ = $\underline{1.735 \times 10^{-6}}$ or $\underline{1.736 \times 10^{-6}}$

The only disadvantage in using natural logs is working with the tables. All the problems you have solved could be done, however, with common or natural logs or for that matter with any log having any number (except 1) as its base, provided you can find the correct log tables. You could convince yourself of this by using natural logs for any or all of the problems in Units 1 to 15. The rules are the same and all "tricks" can be used in both systems.

SELF-TEST

Using the log table, Figure 4, find the natural logs for the following numbers:

1. $\log_e 5781$ = ____________

2. $\log_e 73.55$ = ____________

3. $\log_e 0.7355$ = ____________

4. $\log_e (1.758 \times 10^{18})$ = ____________

5. $\log_e (3.479 \times 10^{-9})$ = ____________

Using the same table, find the following antilogs:

6. antilog 17.5876 = ____________

7. antilog 43.8593 = ____________

8. antilog $\overline{17}.8321$ = ____________

9. antilog 33.7516 = ______________

10. antilog $\overline{23}.7819$ = ______________

Answers

1. $\log_e (5.781 \times 10^3) = 1.7546 + 6.9078$
$= 8.6624$

2. $\log_e (7.355 \times 10^1) = 1.9954 + 2.3026$
$= 4.2980$

3. $\log_e (7.355 \times 10^{-1}) = 1.9954 + \overline{3}.6974$
$= \overline{1}.6928$

4. $\log_e (1.758 \times 10^{18}) = 0.5642 + 41.4466$
$= 42.0108$

5. $\log_e (3.479 \times 10^{-9}) = 1.2468 + \overline{21}.2767$
$= \overline{20}.5235$

6. antilog 16.1181 $= 10^7$
antilog $\underline{1.4695}$ $= 4.347$
antilog 17.5876 $= 4.347 \times 10^7$

7. antilog 43.7492 $= 10^{19}$
antilog $\underline{0.1101}$ $= 1.117$
antilog 43.8593 $= 1.117 \times 10^{19}$

8. antilog $\overline{19}.5793 = 10^{-8}$

antilog $\underline{\ 2.2528\ } = 9.515$

antilog $\overline{17}.8321 = 9.515 \times 10^{-8}$

9. antilog $32.2362 = 10^{14}$

antilog $\underline{\ 1.5154\ } = 4.551$ or 4.552

antilog $33.7516 = 4.551 \times 10^{14}$ or 4.552×10^{14}

10. antilog $\overline{24}.9741 = 10^{-10}$

antilog $\underline{\ 0.8078\ } = 2.243$

antilog $\overline{23}.7819 = 2.243 \times 10^{-10}$

Final Self-Test

Solve the problems below, using common logs. Answers are given on p. 199 . If you wish to solve these problems by using natural logs, solutions with $\log_e$ are given on p.

1. $(385.6)\ (785.9) =$ ____________

2. $\dfrac{1.859}{386.5} =$ ____________

3. $\dfrac{625.9}{.3995} =$ ____________

4. $(.08543)\ (1.965 \times 10^{-6}) =$ ____________

5. $\dfrac{5.853 \times 10^{11}}{6.751 \times 10^{5}} =$ ____________

6. $\dfrac{38.75}{77.5}\ (85.83)\ (576.5) =$ ____________

7. $\frac{1}{317.5}(710.9)(841.6) =$ ________

8. $\frac{(796.5)(817.5)(76.16)}{(593.5)(871.9)(716.9)} =$ ________

9. $\frac{(.8765)(.00375)(.1765)}{(58.5)(.1785)(69.5)} =$ ________

10. $\frac{(8.753 \times 10^{-5})(7.583 \times 10^{4})}{(3.756 \times 10^{-6})(4.565 \times 10^{3})} =$ ________

11. $\frac{(78.75)^{5}\ (231.5)^{\frac{1}{6}}}{\sqrt{785}\ (534.7)^{3}} =$ ________

12. $\frac{\sqrt[5]{891.3}\ \sqrt[7]{37.5}}{\sqrt{476.7}\ (817.5)^{3}} =$ ________

13. $\frac{(896.5)\ \sqrt[4]{.135}\ (81)^{9}}{\sqrt[3]{581.3}\ (796.5)\ (.756)^{\frac{1}{5}}} =$ ________

14. $\frac{(339.8)^{\frac{7}{8}}\ \sqrt[3]{(519)^{2}}\ (354.6)}{(945.6)^{\frac{2}{3}}\ (175.9)\ \sqrt[7]{(256.9)^{3}}} =$ ________

15. $\frac{(.3178)\ (.7195)^{\frac{3}{2}}\ \sqrt[3]{(566)^{5}}}{(.0378)^{\frac{1}{7}}\ \sqrt[3]{(.5773)^{2}}\ (.0375)} =$ ________

Answers Using $\log_{10}$

1. $\log 385.6 + \log 785.9 = 2.5862 + 2.8954$
   $= 5.4816$
   antilog $= 5.4816 = 3.033 \times 10^{5}$

2. $\log 1.859 - \log 386.5 = .2693 - 2.5872$
   $= \bar{3}.6821$
   antilog $= \bar{3}.6821 = 4.81 \times 10^{-3}$

3. $\log 625.9 - \log .3995 = 2.7965 - \bar{1}.6015$
   $= 3.1950$
   antilog $= 3.1950 = 1.567 \times 10^{3}$
   $= 1{,}567$

4. $\log .08543 + \log (1.965 \times 10^{-6}) = \bar{2}.9317 + \bar{6}.2934$
   $= \bar{7}.2251$
   antilog $\bar{7}.2251 = 1.68 \times 10^{-7}$

5. $\log (5.853 \times 10^{11}) - \log (6.751 \times 10^{5}) = 11.7674 - 5.8294$
   $= 5.9380$
   antilog $= 5.9380 = 8.67 \times 10^{5}$

6. $\log 38.75 + \log 85.83 + \log 576.5 + \text{colog } 77.5 = 1.5883 + 1.9337 + 2.7608 + \bar{2}.1107$
   $= 4.3935$
   antilog $4.3935 = 2.474 \times 10^{4}$ or 2.475×10^{4}

7. $\log 710.9 + \log 841.6 + \text{colog } 317.5 = 2.8518 + 2.9251 + \bar{3}.4982$

$= 3.2751$

antilog $3.2751 = 1.884 \times 10^{3}$

$= 1{,}884$

8.

log 796.5	=	2.9012
log 817.5	=	2.9125
log 76.16	=	1.8817
colog 593.5	=	$\bar{3}.2265$
colog 871.9	=	$\bar{3}.0596$
colog 716.9	=	$\bar{3}.1446$
		$\bar{1}.1261$

antilog $\bar{1}.1261 = 1.337 \times 10^{-1} = .1337$

9.

log .8765	=	$\bar{1}.9427$	=	9.9427-10
log .00375	=	$\bar{3}.5740$	=	7.5740-10
log .1765	=	$\bar{1}.2467$	=	9.2467-10
colog 58.5	=	$\bar{2}.2328$	=	8.2328-10
colog .1785	=	.7484	=	.7484
colog 69.5	=	$\bar{2}.1580$	=	8.1580-10
				43.9026-50
	=	$\bar{7}.9026$		

antilog $\bar{7}.9026 = 7.991 \times 10^{-7}$ or 7.992×10^{-7}

10. $\log(8.753 \times 10^{-5}) = \bar{5}.9421 = 5.9421-10$

$\log(7.583 \times 10^{4}) = 4.8799 = 4.8799$

$\text{colog}(3.756 \times 10^{-6}) = 5.4253 = 5.4253$

$\text{colog}(4.565 \times 10^{3}) = \bar{4}.3405 = \underline{6.3405-10}$

$22.5878-20$

$= 2.5878$

$\text{antilog}\ 2.5878 = 387.1$

11. $5 \log 78.75 = 1.8963 \times 5 = 9.4815$

$\frac{1}{6} \log 231.5 = 2.3645 \div 6 = .3941$

$\frac{1}{2}\ \text{colog}\ 785 = \bar{3}.1051 \div 2 = \bar{2}.5526$

$3\ \text{colog}\ 534.7 = \bar{3}.2719 \times 3 = \underline{\bar{9}.8157}$

$.2439$

$\text{antilog}\ .2439 = 1.754$

12. $\frac{1}{5} \log 891.3 = 2.9500 \div 5 = .5900$

$\frac{1}{7} \log 37.5 = 1.5740 \div 7 = .2249$

$\frac{1}{2}\ \text{colog}\ 476.7 = \bar{3}.3218 \div 2 = \bar{2}.6609$

$3\ \text{colog}\ 817.5 = \bar{3}.0875 \times 3 = \underline{\bar{9}.2625}$

$\overline{10}.7383$

$\text{antilog}\ \overline{10}.7383 = 5.474 \times 10^{-10}$

13.

log 896.5	=			2.9526
$\frac{1}{4}$ log 135	=	$\bar{1}.1303 \div 4$	=	$\bar{1}.7826$
9 log 81	=	1.9085 x 9	=	17.1765
$\frac{1}{3}$ colog 581.3	=	$\bar{3}.2358 \div 3$	=	$\bar{1}.0786$
colog 796.5	=			$\bar{3}.0988$
$\frac{1}{5}$ colog .756	=	$.1215 \div 5$	=	.0243
				16.1134
antilog 16.1134	=	1.298×10^{16}		

14.

$\frac{7}{8}$ log 339.8	=	$2.5312 \times \frac{7}{8}$	=	2.2148
$\frac{2}{3}$ log 519	=	$2.7152 \times \frac{2}{3}$	=	1.8101
log 354.6	=		=	2.5497
$\frac{2}{3}$ colog 945.6	=	$\bar{3}.0143 \times \frac{2}{3}$	=	$\bar{2}.0162$
colog 175.9	=			$\bar{3}.7548$
$\frac{3}{7}$ colog 256.9	=	$\bar{3}.5903 \times \frac{3}{7}$	=	$\bar{2}.9673$
				1.3129
antilog 1.3129	=	20.55		

15. $\log .3178 = \qquad = \bar{1}.5022$

$\frac{3}{2} \log .7195 = \bar{1}.8570 \times \frac{3}{2} = \bar{1}.7855$

$\frac{5}{3} \log 566 = 2.7528 \times \frac{5}{3} = 4.5880$

$\frac{1}{7} \text{colog } .0378 = 1.4225 \div 7 = .2032$

$\frac{2}{3} \text{colog } .5773 = .2386 \times \frac{2}{3} = .1591$

$\text{colog } .0375 = \qquad \underline{1.4260}$

5.6640

$\text{antilog } 5.6640 = 4.613 \times 10^5$

Answers Using $\log_e$

1. $\log 385.6 + \log 785.9 = \log (3.856 \times 10^2) + \log (7.859 \times 10^2)$

$= 1.3497 + 4.6052 + 2.0616 + 4.6052$

$= 12.6217$

Alternate method $= \log \left[(385.6)(785.9)\right]$

$= \log \left[3.856 \times 10^2 \times 7.859 + 10^2\right]$

$= \log \left[3.856 \times 7.859 \times 10^4\right]$ *

$= \log 3.856 + \log 7.859 + \log 10^4$ *

* These steps can be omitted, but are shown here for greater clarity. The next examples will not show these steps.

$= 1.3497 + 2.0616 + 9.2103$

$= 12.6216$

antilog 11.5129 $= 10^5$

antilog $\underline{1.1087}$ $= 3.030$

antilog 12.6216 $= 3.030 \times 10^5$

2. log 1.859 + colog 386.5 $=$ log 1.859 + colog (3.865×10^2)

$= 0.6201 + \bar{2}.6480 + \bar{5}.3948$

$= \bar{6}.6629$

antilog $\bar{7}.0922$ $= 10^{-3}$

antilog $\underline{1.5707}$ $= 4.81$

antilog $\bar{6}.6629$ $= 4.81 \times 10^{-3} = 0.00481$

3. log (6.259×10^2) - log (3.995×10^{-1}) $=$ log 6.259 + log 10^3

\- log 3.995*

$= 1.8340 + 6.9078$

$- 1.3851$

$= 7.3567$

antilog 6.9078 $= 10^3$

antilog $\underline{0.4489}$ $= 1.566$ or 1.567

antilog 7.3567 $= 1.566 \times 10^3 = 1566$

$= 1.567 \times 10^3 = 1567$

---

*Powers of 10 can be added or subtracted by inspection if there are only a few in calculation.

---

4. $\log (8.543 \times 10^{-2}) + \log (1.965 \times 10^{-6}) = \log 8.543 + \log 1.965$
$+ \log 10^{-8}$
$= 2.1452 + 0.6755 + \overline{19}.5793$
$= \overline{16}.4000$

antilog $\overline{17}.8819$	=	10^{-7}
antilog $\underline{0.5181}$	=	1.68
antilog $\overline{16}.4000$	=	1.679×10^{-7}

5. $\log (5.853 \times 10^{11}) - \log (6.751 \times 10^{5}) = \log 5.853 - \log 6.751$
$+ \log 10^{6}$
$= 1.7669 - 1.9096 + 13.8155$
$= 13.6728$

antilog 11.5129	=	10^{5}
antilog $\underline{2.1599}$	=	8.67
antilog 13.6728	=	8.67×10^{5}

6. $\log (3.875 \times 10^{1}) + \log (8.583 \times 10^{1}) + \log (5.765 \times 10^{2})$
$+ \text{colog} (7.75 \times 10^{1}) = \log 3.875 + \log 8.583 + \log 5.765$
$+ \text{colog } 7.75 + \log (10^{1} \times 10^{1} \times 10^{2} \times 10^{-1})$

log 3.875	=	1.3546
log 8.583	=	2.1498
log 5.765	=	1.7518
colog 7.75	=	$\overline{3}.9523$
log 10^{3}	=	$\underline{6.9078}$
		10.1163

antilog 9.2103 = 10^4

antilog $\underline{0.9060}$ = 2.474 or 2.475

antilog 10.1163 = 2.474×10^4 or 2.475×10^4

7. $\log (7.109 \times 10^2) + \log (8.416 \times 10^2) - \log (3.175 \times 10^2)$

$= \log 7.109 + \log 8.416 + \text{colog } 3.175 + \log 10^2$

$= 1.9614 + 2.1301 + \bar{2}.8447 + 4.6052 = 7.5414$

antilog 6.9078 = 10^3

antilog $\underline{0.6336}$ = 1.884

antilog 7.5414 = 1884×10^3 = 1,884

8. Simplify

$$\frac{7.965 \times \cancel{10^2} \times 8.175 \times \cancel{10^2} \times 7.616 \times \cancel{10^1}}{5.935 \times \cancel{10^2} \times 8.719 \times \cancel{10^2} \times 7.169 \times \cancel{10^2}_{\;10^1}}$$

log 7.965 = 2.0750

log 8.175 = 2.1011

log 7.616 = 2.0303

colog 5.935 = $\bar{2}.2192$

colog 8.719 = $\bar{3}.8345$

colog 7.169 = $\bar{2}.0302$

colog 10^1 = log 10^{-1} = $\underline{\bar{3}.6974}$

$\bar{3}.9877$

antilog $\bar{3}.6974$ = 10^{-1}

antilog $\underline{0.2903}$ = 1.338

antilog $\bar{3}.9877$ = 1.338×10^{-1} = 0.1338

9. Simplify $$\frac{8.765 \times \cancel{10^{-1}} \times 3.75 \times 10^{-3} \times 1.765 \times 10^{-1}}{5.85 \times 10^{1} \times 1.785 \times \cancel{10^{-1}} \times 6.95 \times 10^{1}}$$

Collect remaining 10^n: $$\frac{10^{-3} \times 10^{-1}}{10^{1} \times 10^{1}} = 10^{-6}$$

$$\frac{(8.765)\ (3.75)\ (1.765)\ (10^{-6})}{(5.85)\ (1.785)\ (6.95)}$$

log 8.765	=	2.1708
log 3.75	=	1.3218
log 1.765	=	0.5682
log 10^{-6}	=	$\overline{14}.1845$
colog 5.85	=	$\overline{2}.2336$
colog 1.785	=	$\overline{1}.4205$
colog 6.95	=	$\overline{2}.0613$
		$\overline{15}.9607$

antilog $\overline{17}.8819$	=	10^{-7}
antilog 2.0788	=	7.995
antilog $\overline{15}.9607$	=	7.995×10^{-7}

10. Simplify: $$\frac{8.753 \times \cancel{10^{-5}} \times 7.583 \times \cancel{10^{4}}}{3.756 \times \cancel{10^{-6}} \times 4.565 \times \cancel{10^{3}}}$$

Collect 10^n: $10^{-5} \times 10^{4} \times 10^{6} \times 10^{-3} = 10^{2}$

log 8.753 = 2.1694

log 7.583 = 2.0259

log 10^2 = 4.6052

colog 3.756 = $\bar{2}.6766$

colog 4.565 = $\underline{\bar{2}.4816}$

5.9587

antilog 4.6052 = 10^2

antilog $\underline{1.3535}$ = 3.871

antilog 5.9587 = 3.871×10^2 = 387.1

11. Simplify $$\frac{\left[(7.875)^5 \times \cancel{10^{1\times 5}}\right]\left[(2.315)^{\frac{1}{6}} \times \cancel{10^{2\times 6}}^{\frac{1}{6}}\right] \times 10^{-\frac{5}{3}}}{\left[(7.85)^{\frac{1}{2}} \times \cancel{10^{2\times 2}}^{\frac{1}{2}}\right]\left[(5.347)^3 \cancel{\times 10^{2\times 3}}\right]}$$

Collect 10^n: $10^5 \times 10^{\frac{1}{3}} \times 10^{-1} \times 10^{-6} = 10^{-\frac{5}{3}}$

5 log 7.875 = 2.0637×5 = 10.3185

$\frac{1}{6}$ log 2.315 = $0.8393 \div 6$ = 0.1399

$\frac{1}{3}$ log 10^{-5} = $\overline{12}.4871 \div 3$ = $\bar{4}.1624$

$\frac{1}{2}$ colog 7.85 = $\bar{3}.9395 \div 2$ = $\bar{2}.9698$

3 colog 5.347 = $\bar{2}.3235 \times 3$ = $\underline{\bar{6}.9705}$

.5611

antilog .5611 = 1.753

12. Simplify $$\frac{\left[(8.913)^{\frac{1}{5}} \times \cancel{10^{2 \times \frac{1}{5}}}\right]\left[(3.75)^{\frac{1}{7}} \times \cancel{10^{1 \times \frac{1}{7}}}\right]}{\left[(4.767)^{\frac{1}{2}} \times \cancel{10^{2 \times \frac{1}{2}}}\right]\left[(8.175)^{3} \times \cancel{10^{2 \times 3}}\right]}$$

Collect 10^n: $10^{\frac{2}{5}} \times 10^{\frac{1}{7}} \times 10^{-1} \times 10^{-6} = 10^{-7} \times 10^{\frac{2}{5}} \times 10^{\frac{1}{7}}$

$\frac{1}{5}$ log 8.913	=	2.1875 ÷ 5	=	0.4375
$\frac{1}{7}$ log 3.75	=	1.3218 ÷ 7	=	0.1888
$\frac{1}{2}$ colog 4.767	=	$\overline{2}$.4383 ÷ 2	=	$\overline{1}$.2192
3 colog 8.175	=	$\overline{3}$.8989 x 2	=	$\overline{7}$.6967
log 10^{-7}	=			$\overline{17}$.8819
$\frac{1}{5}$ log 10^2	=	4.6052 ÷ 5	=	0.9210
$\frac{1}{7}$ log 10	=	2.3026 ÷ 7	=	0.3289
				$\overline{22}$.6740

antilog $\overline{24}$.9741 = 10^{10}

antilog 1.6999 = 5.473 or 5.474

antilog $\overline{22}$.6740 = 5.473×10^{-10} or 5.474×10^{-10}

13. Simplify
$$\frac{\left[8.965 \times \cancel{10^{2}}\right]\left[(1.35)^{\frac{1}{4}} \times \cancel{10^{-1 \times \frac{1}{4}}}\right]\left[(8.1)^{9} \times \cancel{10^{1 \times 9}}\right]}{\left[(5.813)^{\frac{1}{3}} \times \cancel{10^{2 \times \frac{1}{3}}}\right]\left[7.965 \times \cancel{10^{2}}\right]\left[(7.56)^{\frac{1}{5}} \times \cancel{10^{-1 \times \frac{1}{5}}}\right]}$$

Collect 10^n: $10^{2} \times 10^{-\frac{1}{4}} \times 10^{9} \times 10^{-\frac{2}{3}} \times 10^{-2} \times 10^{\frac{1}{5}}$

$= 10^{9} \times 10^{\frac{1}{5}} \times 10^{-\frac{1}{4}} \times 10^{-\frac{1}{5}}$

log 8.965	=			2.1934
$\frac{1}{4}$ log 1.35	=	0.3001 ÷ 4	=	0.0750
9 log 8.1	=	2.0919 x 9	=	18.8271
log 10^{9}	=			20.7233
$\frac{1}{5}$ log 10^{1}	=	2.3026 ÷ 5	=	0.4605
$\frac{1}{4}$ log 10^{-1}	=	$\bar{3}.6974$ ÷ 4	=	$\bar{1}.4244$
$\frac{1}{3}$ log 10^{-2}	=	$\bar{5}.3948$ ÷ 3	=	$\bar{2}.4649$
$\frac{1}{3}$ colog 5.813	=	$\bar{2}.2399$ ÷ 3	=	$\bar{1}.4133$
colog 7.965	=			$\bar{3}.9250$
$\frac{1}{5}$ colog 7.56	=	$\bar{3}.9771$ ÷ 5	=	$\bar{1}.5954$
				37.1023

antilog 36.8414 = 10^{16}

antilog 0.2609 = 1.298

antilog 37.1023 = 1.298 x 10^{16}

14. Simplify $$\frac{\left[(3.398)^{\frac{7}{8}} \cancel{\times 10^{2 \times \frac{7}{8}}}\right]\left[(5.19)^{\frac{2}{3}} \cancel{\times 10^{2 \times \frac{2}{3}}}\right]\left[3.546 \cancel{\times 10^{2}}\right]}{\left[(9.456)^{\frac{2}{3}} \cancel{\times 10^{2 \times \frac{2}{3}}}\right]\left[1.759 \cancel{\times 10^{2}}\right]\left[(2.569)^{\frac{3}{7}} \cancel{\times 10^{2 \times \frac{3}{7}}}\right]}$$

Collect 10^n: $10^{\frac{7}{4}} \times 10^{\frac{4}{3}} \times 10^{2} \times 10^{-\frac{4}{3}} \times 10^{-2} \times 10^{-\frac{6}{7}}$

$\frac{7}{8} \log 3.398 \quad = \quad 1.2232 \times \frac{7}{8} \quad = \quad 1.0703$

$\frac{2}{3} \log 5.19 \quad = \quad 1.6467 \times \frac{2}{3} \quad = \quad 1.0978$

$\log 3.546 \quad = \quad 1.2658$

$\frac{1}{4} \log 10^{7} \quad = \quad 16.1181 \div 4 \quad = \quad 4.0295$

$\frac{1}{7} \log 10^{-6} \quad = \quad \overline{14}.1845 \div 7 \quad = \quad \overline{2}.0264$

$\frac{2}{3} \operatorname{colog} 9.456 \quad = \quad \overline{3}.7534 \times \frac{2}{3} \quad = \quad \overline{2}.5023$

$\operatorname{colog} 1.759 \quad = \quad \overline{1}.4353$

$\frac{3}{7} \operatorname{colog} 2.569 \quad = \quad \overline{1}.0565 \times \frac{3}{7} \quad = \quad \underline{\overline{1}.5956}$

3.0230

antilog $2.3026 \quad = \quad 10^{1}$

antilog $\underline{0.7204} \quad = \quad 2.055$

antilog $3.0230 \quad = \quad 2.055 \times 10^{1} \quad = \quad 20.55$

15. Simplify $$\frac{\left[3.178 \times \cancel{10^{-1}}\right]\left[(7.195)^{\frac{3}{2}} \times \cancel{10^{-1 \times \frac{3}{2}}}\right]\left[(5.66)^{\frac{5}{3}} \times \cancel{10^{2 \times \frac{5}{3}}}\right]}{\left[(3.78)^{\frac{1}{7}} \times \cancel{10^{-2 \times \frac{1}{7}}}\right]\left[(5.773)^{\frac{2}{3}} \times \cancel{10^{-1 \times \frac{2}{3}}}\right]\left[3.75 \times \cancel{10^{-2}}\right]}$$

Collect 10^n: $10^{-1} \times 10^{-\frac{3}{2}} \times 10^{\frac{10}{3} + \frac{2}{3}} \times 10^{\frac{2}{7}} \times 10^{\frac{2}{3}} \times 10^{2} = 10^{-\frac{1}{2}}$

$\times\ 10^{4} \times 10^{\frac{2}{7}}$

$\log 3.178$	=			1.1562
$\frac{3}{2}\log 7.195$	=	$1.9734 \times \frac{3}{2}$	=	2.9601
$\frac{5}{3}\log 5.66$	=	$1.7334 \times \frac{5}{3}$	=	2.8890
$\frac{1}{2}\log 10^{-1}$	=	$\bar{3}.6974 \div 2$	=	$\bar{2}.8487$
$\log 10^{4}$	=			9.2103
$\frac{1}{7}\log 10^{2}$	=	$4.6052 \div 7$	=	0.6579
$\frac{1}{7}$ colog 3.78	=	$\bar{2}.6703 \div 7$	=	$\bar{1}.8100$
$\frac{2}{3}$ colog 5.773	=	$\bar{2}.2468 \times \frac{2}{3}$	=	$\bar{2}.8312$
colog 3.75	=		=	$\underline{\bar{2}.6782}$
				13.0416

antilog 11.5129 = 10^{5}

antilog $\underline{\ 1.5287\ }$ = 4.612 or 4.613

antilog 13.0416 = 4.612×10^{5} or 4.613×10^{5}

Common logarithms and proportional parts

N	0	1	2	3	4	5	6	7	8	9	1	2	3	4	5	6	7	8	9
											proportional parts								
10	0000	0043	0086	0128	0170	0212	0253	0294	0334	0374	4	8	12	17	21	25	29	33	37
11	0414	0453	0492	0531	0569	0607	0645	0682	0719	0755	4	8	11	15	19	23	26	30	34
12	0792	0828	0864	0899	0934	0969	1004	1038	1072	1106	3	7	10	14	17	21	24	28	31
13	1139	1173	1206	1239	1271	1303	1335	1367	1399	1430	3	6	10	13	16	19	23	26	29
14	1461	1492	1523	1553	1584	1614	1644	1673	1703	1732	3	6	9	12	15	18	21	24	27
15	1761	1790	1818	1847	1875	1903	1931	1959	1987	2014	3	6	8	11	14	17	20	22	25
16	2041	2068	2095	2122	2148	2175	2201	2227	2253	2279	3	5	8	11	13	16	18	21	24
17	2304	2330	2355	2380	2405	2430	2455	2480	2504	2529	2	5	7	10	12	15	17	20	22
18	2553	2577	2601	2625	2648	2672	2695	2718	2742	2765	2	5	7	9	12	14	16	19	21
19	2788	2810	2833	2856	2878	2900	2923	2945	2967	2989	2	4	7	9	11	13	16	18	20
20	3010	3032	3054	3075	3096	3118	3139	3160	3181	3201	2	4	6	8	11	13	15	17	19
21	3222	3243	3263	3284	3304	3324	3345	3365	3385	3404	2	4	6	8	10	12	14	16	18
22	3424	3444	3464	3483	3502	3522	3541	3560	3579	3598	2	4	6	8	10	12	14	15	17
23	3617	3636	3655	3674	3692	3711	3729	3747	3766	3784	2	4	6	7	9	11	13	15	17
24	3802	3820	3838	3856	3874	3892	3909	3927	3945	3962	2	4	5	7	9	11	12	14	16
25	3979	3997	4014	4031	4048	4065	4082	4099	4116	4133	2	3	5	7	9	10	12	14	15
26	4150	4166	4183	4200	4216	4232	4249	4265	4281	4298	2	3	5	7	8	10	11	13	15
27	4314	4330	4346	4362	4378	4393	4409	4425	4440	4456	2	3	5	6	8	9	11	13	14
28	4472	4487	4502	4518	4533	4548	4564	4579	4594	4609	2	3	5	6	8	9	11	12	14
29	4624	4639	4654	4669	4683	4698	4713	4728	4742	4757	1	3	4	6	7	9	10	12	13
30	4771	4786	4800	4814	4829	4843	4857	4871	4886	4900	1	3	4	6	7	9	10	11	13
31	4914	4928	4942	4955	4969	4983	4997	5011	5024	5038	1	3	4	6	7	8	10	11	12
32	5051	5065	5079	5092	5105	5119	5132	5145	5159	5172	1	3	4	5	7	8	9	11	12
33	5185	5198	5211	5224	5237	5250	5263	5276	5289	5302	1	3	4	5	6	8	9	10	12
34	5315	5328	5340	5353	5366	5378	5391	5403	5416	5428	1	3	4	5	6	8	9	10	11
35	5441	5453	5465	5478	5490	5502	5514	5527	5539	5551	1	2	4	5	6	7	9	10	11
36	5563	5575	5587	5599	5611	5623	5635	5647	5658	5670	1	2	4	5	6	7	8	10	11
37	5682	5694	5705	5717	5729	5740	5752	5763	5775	5786	1	2	3	5	6	7	8	9	10
38	5798	5809	5821	5832	5843	5855	5866	5877	5888	5899	1	2	3	5	6	7	8	9	10
39	5911	5922	5933	5944	5955	5966	5977	5988	5999	6010	1	2	3	4	5	7	8	9	10
40	6021	6031	6042	6053	6064	6075	6085	6096	6107	6117	1	2	3	4	5	6	8	9	10
41	6128	6138	6149	6160	6170	6180	6191	6201	6212	6222	1	2	3	4	5	6	7	8	9
42	6232	6243	6253	6263	6274	6284	6294	6304	6314	6325	1	2	3	4	5	6	7	8	9
43	6335	6345	6355	6365	6375	6385	6395	6405	6415	6425	1	2	3	4	5	6	7	8	9
44	6435	6444	6454	6464	6474	6484	6493	6503	6513	6522	1	2	3	4	5	6	7	8	9
45	6532	6542	6551	6561	6571	6580	6590	6599	6609	6618	1	2	3	4	5	6	7	8	9
46	6628	6637	6646	6656	6665	6675	6684	6693	6702	6712	1	2	3	4	5	6	7	7	8
47	6721	6730	6739	6749	6758	6767	6776	6785	6794	6803	1	2	3	4	5	5	6	7	8
48	6812	6821	6830	6839	6848	6857	6866	6875	6884	6893	1	2	3	4	4	5	6	7	8
49	6902	6911	6920	6928	6937	6946	6955	6964	6972	6981	1	2	3	4	4	5	6	7	8
50	6990	6998	7007	7016	7024	7033	7042	7050	7059	7067	1	2	3	3	4	5	6	7	8
51	7076	7084	7093	7101	7110	7118	7126	7135	7143	7152	1	2	3	3	4	5	6	7	8
52	7160	7168	7177	7185	7193	7202	7210	7218	7226	7235	1	2	2	3	4	5	6	7	7
53	7243	7251	7259	7267	7275	7284	7292	7300	7308	7316	1	2	2	3	4	5	6	6	7
54	7324	7332	7340	7348	7356	7364	7372	7380	7388	7396	1	2	2	3	4	5	6	6	7

N	0	1	2	3	4	5	6	7	8	9	1	2	3	4	5	6	7	8	9
											proportional parts								
55	7404	7412	7419	7427	7435	7443	7451	7459	7466	7474	1	2	2	3	4	5	5	6	7
56	7482	7490	7497	7505	7513	7520	7528	7536	7543	7551	1	2	2	3	4	5	5	6	7
57	7559	7566	7574	7582	7589	7597	7604	7612	7619	7627	1	2	2	3	4	5	5	6	7
58	7634	7642	7649	7657	7664	7672	7679	7686	7694	7701	1	1	2	3	4	4	5	6	7
59	7709	7716	7723	7731	7738	7745	7752	7760	7767	7774	1	1	2	3	4	4	5	6	7
60	7782	7789	7796	7803	7810	7818	7825	7832	7839	7846	1	1	2	3	4	4	5	6	6
61	7853	7860	7868	7875	7882	7889	7896	7903	7910	7917	1	1	2	3	4	4	5	6	6
62	7924	7931	7938	7945	7952	7959	7966	7973	7980	7987	1	1	2	3	3	4	5	6	6
63	7993	8000	8007	8014	8021	8028	8035	8041	8048	8055	1	1	2	3	3	4	5	5	6
64	8062	8069	8075	8082	8089	8096	8102	8109	8116	8122	1	1	2	3	3	4	5	5	6
65	8129	8136	8142	8149	8156	8162	8169	8176	8182	8189	1	1	2	3	3	4	5	5	6
66	8195	8202	8209	8215	8222	8228	8235	8241	8248	8254	1	1	2	3	3	4	5	5	6
67	8261	8267	8274	8280	8287	8293	8299	8306	8312	8319	1	1	2	3	3	4	5	5	6
68	8325	8331	8338	8344	8351	8357	8363	8370	8376	8382	1	1	2	3	3	4	4	5	6
69	8388	8395	8401	8407	8414	8420	8426	8432	8439	8445	1	1	2	2	3	4	4	5	6
70	8451	8457	8463	8470	8476	8482	8488	8494	8500	8506	1	1	2	2	3	4	4	5	6
71	8513	8519	8525	8531	8537	8543	8549	8555	8561	8567	1	1	2	2	3	4	4	5	5
72	8573	8579	8585	8591	8597	8603	8609	8615	8621	8627	1	1	2	2	3	4	4	5	5
73	8633	8639	8645	8651	8657	8663	8669	8675	8681	8686	1	1	2	2	3	4	4	5	5
74	8692	8698	8704	8710	8716	8722	8727	8733	8739	8745	1	1	2	2	3	4	4	5	5
75	8751	8756	8762	8768	8774	8779	8785	8791	8797	8802	1	1	2	2	3	3	4	5	5
76	8808	8814	8820	8825	8831	8837	8842	8848	8854	8859	1	1	2	2	3	3	4	5	5
77	8865	8871	8876	8882	8887	8893	8899	8904	8910	8915	1	1	2	2	3	3	4	4	5
78	8921	8927	8932	8938	8943	8949	8954	8960	8965	8971	1	1	2	2	3	3	4	4	5
79	8976	8982	8987	8993	8998	9004	9009	9015	9020	9025	1	1	2	2	3	3	4	4	5
80	9031	9036	9042	9047	9053	9058	9063	9069	9074	9079	1	1	2	2	3	3	4	4	5
81	9085	9090	9096	9101	9106	9112	9117	9122	9128	9133	1	1	2	2	3	3	4	4	5
82	9138	9143	9149	9154	9159	9165	9170	9175	9180	9186	1	1	2	2	3	3	4	4	5
83	9191	9196	9201	9206	9212	9217	9222	9227	9232	9238	1	1	2	2	3	3	4	4	5
84	9243	9248	9253	9258	9263	9269	9274	9279	9284	9289	1	1	2	2	3	3	4	4	5
85	9294	9299	9304	9309	9315	9320	9325	9330	9335	9340	1	1	2	2	3	3	4	4	5
86	9345	9350	9355	9360	9365	9370	9375	9380	9385	9390	1	1	2	2	3	3	4	4	5
87	9395	9400	9405	9410	9415	9420	9425	9430	9435	9440	0	1	1	2	2	3	3	4	4
88	9445	9450	9455	9460	9465	9469	9474	9479	9484	9489	0	1	1	2	2	3	3	4	4
89	9494	9499	9504	9509	9513	9518	9523	9528	9533	9538	0	1	1	2	2	3	3	4	4
90	9542	9547	9552	9557	9562	9566	9571	9576	9581	9586	0	1	1	2	2	3	3	4	4
91	9590	9595	9600	9605	9609	9614	9619	9624	9628	9633	0	1	1	2	2	3	3	4	4
92	9638	9643	9647	9652	9657	9661	9666	9671	9675	9680	0	1	1	2	2	3	3	4	4
93	9685	9689	9694	9699	9703	9708	9713	9717	9722	9727	0	1	1	2	2	3	3	4	4
94	9731	9736	9741	9745	9750	9754	9759	9763	9768	9773	0	1	1	2	2	3	3	4	4
95	9777	9782	9786	9791	9795	9800	9805	9809	9814	9818	0	1	1	2	2	3	3	4	4
96	9823	9827	9832	9836	9841	9845	9850	9854	9859	9863	0	1	1	2	2	3	3	4	4
97	9868	9872	9877	9881	9886	9890	9894	9899	9903	9908	0	1	1	2	2	3	3	4	4
98	9912	9917	9921	9926	9930	9934	9939	9943	9948	9952	0	1	1	2	2	3	3	4	4
99	9956	9961	9965	9969	9974	9978	9983	9987	9991	9996	0	1	1	2	2	3	3	3	4

Natural logarithms

N	0	1	2	3	4	5	6	7	8	9	mean differences 1	2	3	4	5	6	7	8	9
1.0	0.0000	0100	0198	0296	0392	0488	0583	0677	0770	0862	10	19	29	38	48	57	67	76	86
1.1	0.0953	1044	1133	1222	1310	1398	1484	1570	1655	1740	9	17	26	35	44	52	61	70	78
1.2	0.1823	1906	1989	2070	2151	2231	2311	2390	2469	2546	8	16	24	32	40	48	56	64	72
1.3	0.2624	2700	2776	2852	2927	3001	3075	3148	3221	3293	7	15	22	30	37	44	52	59	67
1.4	0.3365	3436	3507	3577	3646	3716	3784	3853	3920	3988	7	14	21	28	35	41	48	55	62
1.5	0.4055	4121	4187	4253	4318	4383	4447	4511	4574	4637	6	13	19	26	32	39	45	52	58
1.6	0.4700	4762	4824	4886	4947	5008	5068	5128	5188	5247	6	12	18	24	30	36	42	48	55
1.7	0.5306	5365	5423	5481	5539	5596	5653	5710	5766	5822	6	11	17	23	29	34	40	46	51
1.8	0.5878	5933	5988	6043	6098	6152	6206	6259	6313	6366	5	11	16	22	27	32	38	43	49
1.9	0.6419	6471	6523	6575	6627	6678	6729	6780	6831	6881	5	10	15	20	26	31	36	41	46
2.0	0.6931	6981	7031	7080	7129	7178	7227	7275	7324	7372	5	10	15	20	24	29	34	39	44
2.1	0.7419	7467	7514	7561	7608	7655	7701	7747	7793	7839	5	9	14	19	23	28	33	37	42
2.2	0.7885	7930	7975	8020	8065	8109	8154	8198	8242	8286	4	9	13	18	22	27	31	36	40
2.3	0.8329	8372	8416	8459	8502	8544	8587	8629	8671	8713	4	9	13	17	21	26	30	34	38
2.4	0.8755	8796	8838	8879	8920	8961	9002	9042	9083	9123	4	8	12	16	20	24	29	33	37
2.5	0.9163	9203	9243	9282	9322	9361	9400	9439	9478	9517	4	8	12	16	20	24	27	31	35
2.6	0.9555	9594	9632	9670	9708	9746	9783	9821	9858	9895	4	8	11	15	19	23	26	30	34
2.7	0.9933	9969	1.0006	0043	0080	0116	0152	0188	0225	0260	4	7	11	15	18	22	25	29	33
2.8	1.0296	0332	0367	0403	0438	0473	0508	0543	0578	0613	4	7	11	14	18	21	25	28	32
2.9	1.0647	0682	0716	0750	0784	0818	0852	0886	0919	0953	3	7	10	14	17	20	24	27	31
3.0	1.0986	1019	1053	1086	1119	1151	1184	1217	1249	1282	3	7	10	13	16	20	23	26	30
3.1	1.1314	1346	1378	1410	1442	1474	1506	1537	1569	1600	3	6	10	13	16	19	22	25	29
3.2	1.1632	1663	1694	1725	1756	1787	1817	1848	1878	1909	3	6	9	12	15	18	22	25	28
3.3	1.1939	1969	2000	2030	2060	2090	2119	2149	2179	2208	3	6	9	12	15	18	21	24	27
3.4	1.2238	2267	2296	2326	2355	2384	2413	2442	2470	2499	3	6	9	12	15	17	20	23	26
3.5	1.2528	2556	2585	2613	2641	2669	2698	2726	2754	2782	3	6	8	11	14	17	20	23	25
3.6	1.2809	2837	2865	2892	2920	2947	2975	3002	3029	3056	3	5	8	11	14	16	19	22	25
3.7	1.3083	3110	3137	3164	3191	3218	3244	3271	3297	3324	3	5	8	11	13	16	19	21	24
3.8	1.3350	3376	3403	3429	3455	3481	3507	3533	3558	3584	3	5	8	10	13	16	18	21	23
3.9	1.3610	3635	3661	3686	3712	3737	3762	3788	3813	3838	3	5	8	10	13	15	18	20	23
4.0	1.3863	3888	3913	3938	3962	3987	4012	4036	4061	4085	2	5	7	10	12	15	17	20	22
4.1	1.4110	4134	4159	4183	4207	4231	4255	4279	4303	4327	2	5	7	10	12	14	17	19	22
4.2	1.4351	4375	4398	4422	4446	4469	4493	4516	4540	4563	2	5	7	9	12	14	16	19	21
4.3	1.4586	4609	4633	4656	4679	4702	4725	4748	4770	4793	2	5	7	9	12	14	16	18	21
4.4	1.4816	4839	4861	4884	4907	4929	4951	4974	4996	5019	2	5	7	9	11	14	16	18	20
4.5	1.5041	5063	5085	5107	5129	5151	5173	5195	5217	5239	2	4	7	9	11	13	15	18	20
4.6	1.5261	5282	5304	5326	5347	5369	5390	5412	5433	5454	2	4	6	9	11	13	15	17	19
4.7	1.5476	5497	5518	5539	5560	5581	5602	5623	5644	5665	2	4	6	8	11	13	15	17	19
4.8	1.5686	5707	5728	5748	5769	5790	5810	5831	5851	5872	2	4	6	8	10	12	14	16	19
4.9	1.5892	5913	5933	5953	5974	5994	6014	6034	6054	6074	2	4	6	8	10	12	14	16	18
5.0	1.6094	6114	6134	6154	6174	6194	6214	6233	6253	6273	2	4	6	8	10	12	14	16	18
5.1	1.6292	6312	6332	6351	6371	6390	6409	6429	6448	6467	2	4	6	8	10	12	14	16	18
5.2	1.6487	6506	6525	6544	6563	6582	6601	6620	6639	6658	2	4	6	8	10	11	13	15	17
5.3	1.6677	6696	6715	6734	6752	6771	6790	6808	6827	6845	2	4	6	7	9	11	13	15	17
5.4	1.6864	6882	6901	6919	6938	6956	6974	6993	7011	7029	2	4	5	7	9	11	13	15	17

Natural logarithms of 10^{+n}

n	1	2	3	4	5	6	7	8	9
$\log_e 10^n$	2.3026	4.6052	6.9078	9.2103	11.5129	13.8155	16.1181	18.4207	20.7233

N	0	1	2	3	4	5	6	7	8	9	mean differences 1	2	3	4	5	6	7	8	9
5.5	1.7047	7066	7084	7102	7120	7138	7156	7174	7192	7210	2	4	5	7	9	11	13	14	16
5.6	1.7228	7246	7263	7281	7299	7317	7334	7352	7370	7387	2	4	5	7	9	11	12	14	16
5.7	1.7405	7422	7440	7457	7475	7492	7509	7527	7544	7561	2	3	5	7	9	10	12	14	16
5.8	1.7579	7596	7613	7630	7647	7664	7681	7699	7716	7733	2	3	5	7	9	10	12	14	15
5.9	1.7750	7766	7783	7800	7817	7834	7851	7867	7884	7901	2	3	5	7	8	10	12	13	15
6.0	1.7918	7934	7951	7967	7984	8001	8017	8034	8050	8066	2	3	5	7	8	10	12	13	15
6.1	1.8083	8099	8116	8132	8148	8165	8181	8197	8213	8229	2	3	5	6	8	10	11	13	15
6.2	1.8245	8262	8278	8294	8310	8326	8342	8358	8374	8390	2	3	5	6	8	10	11	13	14
6.3	1.8405	8421	8437	8453	8469	8485	8500	8516	8532	8547	2	3	5	6	8	9	11	13	14
6.4	1.8563	8579	8594	8610	8625	8641	8656	8672	8687	8703	2	3	5	6	8	9	11	12	14
6.5	1.8718	8733	8749	8764	8779	8795	8810	8825	8840	8856	2	3	5	6	8	9	11	12	14
6.6	1.8871	8886	8901	8916	8931	8946	8961	8976	8991	9006	2	3	5	6	8	9	11	12	14
6.7	1.9021	9036	9051	9066	9081	9095	9110	9125	9140	9155	1	3	4	6	7	9	10	12	13
6.8	1.9169	9184	9199	9213	9228	9242	9257	9272	9286	9301	1	3	4	6	7	9	10	12	13
6.9	1.9315	9330	9344	9359	9373	9387	9402	9416	9430	9445	1	3	4	6	7	9	10	12	13
7.0	1.9459	9473	9488	9502	9516	9530	9544	9559	9573	9587	1	3	4	6	7	9	10	11	13
7.1	1.9601	9615	9629	9643	9657	9671	9685	9699	9713	9727	1	3	4	6	7	8	10	11	13
7.2	1.9741	9755	9769	9782	9796	9810	9824	9838	9851	9865	1	3	4	6	7	8	10	11	12
7.3	1.9879	9892	9906	9920	9933	9947	9961	9974	9988	2.0001	1	3	4	5	7	8	10	11	12
7.4	2.0015	0028	0042	0055	0069	0082	0096	0109	0122	0136	1	3	4	5	7	8	9	11	12
7.5	2.0149	0162	0176	0189	0202	0215	0229	0242	0255	0268	1	3	4	5	7	8	9	11	12
7.6	2.0281	0295	0308	0321	0334	0347	0360	0373	0386	0399	1	3	4	5	7	8	9	10	12
7.7	2.0412	0425	0438	0451	0464	0477	0490	0503	0516	0528	1	3	4	5	6	8	9	10	12
7.8	2.0541	0554	0567	0580	0592	0605	0618	0631	0643	0656	1	3	4	5	6	8	9	10	11
7.9	2.0669	0681	0694	0707	0719	0732	0744	0757	0769	0782	1	3	4	5	6	8	9	10	11
8.0	2.0794	0807	0819	0832	0844	0857	0869	0882	0894	0906	1	3	4	5	6	7	9	10	11
8.1	2.0919	0931	0943	0956	0968	0980	0992	1005	1017	1029	1	2	4	5	6	7	9	10	11
8.2	2.1041	1054	1066	1078	1090	1102	1114	1126	1138	1150	1	2	4	5	6	7	9	10	11
8.3	2.1163	1175	1187	1199	1211	1223	1235	1247	1258	1270	1	2	4	5	6	7	8	10	11
8.4	2.1282	1294	1306	1318	1330	1342	1353	1365	1377	1339	1	2	4	5	6	7	8	9	11
8.5	2.1401	1412	1424	1436	1448	1459	1471	1483	1494	1506	1	2	4	5	6	7	8	9	11
8.6	2.1518	1529	1541	1552	1564	1576	1587	1599	1610	1622	1	2	3	5	6	7	8	9	10
8.7	2.1633	1645	1656	1668	1679	1691	1702	1713	1725	1736	1	2	3	5	6	7	8	9	10
8.8	2.1748	1759	1770	1782	1793	1804	1815	1827	1838	1849	1	2	3	5	6	7	8	9	10
8.9	2.1861	1872	1883	1894	1905	1917	1928	1939	1950	1961	1	2	3	4	6	7	8	9	10
9.0	2.1972	1983	1994	2006	2017	2028	2039	2050	2061	2072	1	2	3	4	6	7	8	9	10
9.1	2.2083	2094	2105	2116	2127	2138	2148	2159	2170	2181	1	2	3	4	5	7	8	9	10
9.2	2.2192	2203	2214	2225	2235	2246	2257	2268	2279	2289	1	2	3	4	5	6	8	9	10
9.3	2.2300	2311	2322	2332	2343	2354	2364	2375	2386	2396	1	2	3	4	5	6	7	9	10
9.4	2.2407	2418	2428	2439	2450	2460	2471	2481	2492	2502	1	2	3	4	5	6	7	8	10
9.5	2.2513	2523	2534	2544	2555	2565	2576	2586	2597	2607	1	2	3	4	5	6	7	8	9
9.6	2.2618	2628	2638	2649	2659	2670	2680	2690	2701	2711	1	2	3	4	5	6	7	8	9
9.7	2.2721	2732	2742	2752	2762	2773	2783	2793	2803	2814	1	2	3	4	5	6	7	8	9
9.8	2.2824	2834	2844	2854	2865	2875	2885	2895	2905	2915	1	2	3	4	5	6	7	8	9
9.9	2.2925	2935	2946	2956	2966	2976	2986	2996	3006	3016	1	2	3	4	5	6	7	8	9
10.0	2.3026																		

Natural logarithms of 10^{-n}

n	1	2	3	4	5	6	7	8	9
$\log_e 10^{-n}$	$\overline{3}.6974$	$\overline{5}.3948$	$\overline{7}.0922$	$\overline{10}.7897$	$\overline{12}.4871$	$\overline{14}.1845$	$\overline{17}.8819$	$\overline{19}.5793$	$\overline{21}.2767$